Make Way for the Superhumans

'One of the most thoughtful meditations on the future that I have read. … the book is carefully and conscientiously crafted, and meticulously argued. Bess is also impartial, giving a fair hearing to contradictory arguments, and wrestling fairly with the ideas as he encounters them.'

Institute for Ethics and Emerging Technologies

'Rejuvenation therapies that could potentially extend human lifespans to 160 years or more, chemical or bioelectronic cognitive enhancement that could double or triple IQ scores, bioelectronic devices for modulating brain processes including "pleasure centres", so-called "designer babies", and much more are poised to cross the threshhold from science fiction to reality in the near future. Michael Bess offers a sober prediction of how such advances will directly affect human society, and the ethical dilemmas that could result. … fascinating from cover to cover and near-impossible to put down. Highly recommended!'

Midwest Book Review

MAKE WAY FOR THE

SUPER HUMANS

ALSO BY MICHAEL BESS

Choices Under Fire:
Moral Dimensions of World War II

The Light-Green Society:
Ecology and Technological Modernity in France, 1960–2000

Realism, Utopia, and the Mushroom Cloud:
Four Activist Intellectuals and Their Strategies for Peace, 1945–1989

MAKE WAY FOR THE

SUPER HUMANS

*How the science of bio-enhancement
is transforming our world, and
how we need to deal with it*

MICHAEL BESS

ICON

This edition published in the UK in 2016
by Icon Books Ltd, Omnibus Business Centre,
39–41 North Road, London N7 9DP
email: info@iconbooks.com
www.iconbooks.com

First published in the USA in 2015
under the title *Our Grandchildren Redesigned*
by Beacon Press Books, under the auspices of
the Unitarian Universalist Association of Congregations

Sold in the UK, Europe and Asia
by Faber & Faber Ltd, Bloomsbury House,
74–77 Great Russell Street,
London WC1B 3DA or their agents

Distributed in the UK, Europe and Asia
by Grantham Book Services, Trent Road,
Grantham NG31 7XQ

Distributed in Australia and New Zealand
by Allen & Unwin Pty Ltd,
PO Box 8500, 83 Alexander Street,
Crows Nest, NSW 2065

Distributed in South Africa
by Jonathan Ball, Office B4, The District,
41 Sir Lowry Road, Woodstock 7925

Distributed in India by Penguin Books India,
7th Floor, Infinity Tower – C, DLF Cyber City,
Gurgaon 122002, Haryana

ISBN: 978-178578-101-8

Text design and composition by Kim Arney

Printed and bound in the UK by Clays Ltd, St Ives plc

In the place that is my own place, whose earth
I am shaped in and must bear, there is an old tree growing,
a great sycamore that is a wondrous healer of itself.
Fences have been tied to it, nails driven into it,
hacks and whittles cut in it, the lightning has burned it.
There is no year it has flourished in
that has not harmed it. There is a hollow in it
that is its death, though its living brims whitely
at the lip of the darkness and flows outward.
Over all its scars has come the seamless white
of the bark. It bears the gnarls of its history
healed over. It has risen to a strange perfection
in the warp and bending of its long growth.
It has gathered all accidents into its purpose.
It has become the intention and radiance of its dark fate.
It is a fact, sublime, mystical and unassailable.
In all the country there is no other like it.
I recognize in it a principle, an indwelling
the same as itself, and greater, that I would be ruled by.
I see that it stands in its place and feeds upon it,
and is fed upon, and is native, and maker.

—WENDELL BERRY,
"The Sycamore" (1998)[1]

CONTENTS

ONLINE COMPANION WEBSITE TO THIS BOOK:

www.ourgrandchildrenredesigned.org

Updates on Science and Technology

Updates on Social and Cultural Implications

Appendices

Dialogue Page

Full Bibliography

Welcome to the Future

Reflections from an autumn afternoon in 2049

- They've nicknamed me Mr. Amish. That's what they're calling people like me these days. I smile, I shrug. My son, my grandkids—they're not being mean. They put their arm around my shoulders, squeezing gently, I can feel the affection. Amish grandpa. It's kind of like I'm a living window into the past.

  Ever since Martha died, nineteen years ago, I stopped doing all the fancy pills and implants and epigenetic boosts. I kept the bioenhancements I had, and left it at that. No more upgrades, no more tweaks. I'd had enough.

  It's hard for them to understand. They argue with me, my son Pete pleads with me. "You'll feel so much better, Dad. You know you will."

  But what if I don't want to feel better?

  I glance across the living room at my grandson Kenny, playing 3-D chess with his younger sister Gwendolyn. He's ten, she's eight. They're sitting across from each other on the floor near the window, eyes closed, wearing their headsets. The old maple stands just outside, its foliage of orange and yellow raining down color with the afternoon sunlight.

  I used to be really good at chess—the old-fashioned 2-D kind. Kenny started beating me at that when he was six. Now it's 3-D chess they all play, the moves and threats and gambits simultaneously above and below, exponentially more complicated. Gwen's giving him a run for his money. She's even better at it than he is.

  Makes you wonder, though. We give them neuroceuticals with their vitamins in the morning, we make them wear cortical stimulation headsets while they do their homework, we epigenetically tweak their memory and acuity. Then my son acts surprised when he finds his daughter crying in her bedroom, reading the *New York Times*. How do you explain genocide to an eight-year-old kid? Problem is, she understands the article all too well. It's human nature that has her baffled.

  When I bring this up with Pete, he just looks at me like I'm clueless. "What do you want me to do, Dad? Bring everything to a halt? Have everyone else in their grade running circles around 'em?"

I'm not the only one who feels this way, of course. All of us who were born before the onset of it all, we sometimes have a hard time adapting.

I just never thought it would happen so fast, so soon. I was born in 1979. I witnessed the birth of the Web, watched it spread through our culture. I understood the acceleration of technology. At least I thought I did. Hell, I even invested in it, made a pile of money in epigenetics stock.

So many aspects of it are amazing—the inventions unfolding all around us like one of those fast-forward movies of a garden in bloom. They've improved our lives in so many ways. I'm turning seventy this December, and I feel like I did when I was forty-five. Better, actually.

But other aspects of it—if I'm really honest . . .

■   ■   ■

Over the coming decades—probably a lot sooner than most people realize—the next great wave of technological change will wash over our lives. Its impact will be similar in sweep and rapidity to the advent of computers, cell phones, and the Web; but this time around, it is not our gadgets that will be transformed—it is we ourselves, our bodies, our minds. This will be a shift that cuts even more deeply than the great industrial revolutions of the past. It will not only alter how we make a living, communicate, and interact with each other, but will offer direct and precise control over our own physical and mental states. People will be able to sculpt their own selfhood over time, reshaping their bodies, augmenting their cognition, reconfiguring their character and personality. We will live through this process, year by year, marveling all the while at how malleable our species turned out to be.

If you talk to the authors of this revolution—the scientists, doctors, and engineers who labor tirelessly at the vanguard of biotechnology—most of them will deny that this is what they have in mind. They are not seeking to bring about the transmogrification of the human species, they insist: they are simply doing their best to heal the sick, to repair the injured. But once you stand back and look at the big picture, sizing up the cumulative impact of all their brilliant efforts, a different conclusion emerges. Whether they intend it or not, they are giving our species the instruments with which to radically redesign itself. Those instruments are already becoming available in crude form today, and they will fully come into their own over the next few decades. By the time our grandchildren have grown to adulthood, this wave of change will have passed through our civilization.

The results will be mixed. Some of the new bioenhanced capabilities will be splendid to behold (and to experience). People will live longer, healthier, more productive lives; they will connect with each other in seamless webs of direct interactivity; they will be able to fine-tune their own moods and thought processes; they will interact with machines in entirely new ways; their augmented minds will generate staggeringly complex and subtle forms of knowledge and insight.

At the same time, these technologies will also create formidable challenges. If only the rich have access to the most potent bioenhancements, this will exacerbate the already grievous rift between haves and have-nots. Competition will be keen for the most sophisticated enhancement products—because an individual's professional and social success will be at stake. As these technologies advance, they will continuously raise the bar of "normal" performance, forcing people to engage in constant cycles of upgrades and boosts merely to keep up. People will tend to identify strongly with their particular "enhancement profiles," clustering together in novel social and cultural groupings that could lead to new forms of prejudice, rivalry, and outright conflict. Some bioenhancements will offer such fine-grained control over feelings and moods that they risk turning people into emotional puppets. Individuals who boost their traits beyond a certain threshold may acquire such extreme capabilities that they will no longer be recognized as unambiguously human.

Until recently in human history, the major technological watersheds all came about incrementally, spread out over centuries or longer. Think, for example, of the shift from stone to metal tools, the transition from nomadic hunter-gathering to settled agriculture, or the substitution of mechanical power for human and animal sources of energy. In all these cases, people and social systems had time to adapt: they gradually developed new values and habits to accommodate the transformed material conditions. But this is not the case with the current epochal shift. This time around, the radical innovations are coming upon us with relative suddenness—in a time frame that encompasses four or five decades, a century at most. A central argument of this book is that contemporary society is dangerously unprepared for the dramatic changes it is about to experience, down this road on which it is already advancing at an accelerating pace.

More specifically, my thesis is threefold.

1. *It's almost certainly going to happen.* Advanced enhancement technologies will become a reality, even if they are initially opposed by significant minorities of the population. They will come into being, not because of

some impersonal, deterministic process outside of human control, but rather because people will actively choose them. Many of these inventions will prove irresistibly desirable for a sufficiently large number of citizens, who will eagerly purchase them once they are declared safe and made available on the market.

2. *It will bring both opportunity and peril.* Most bioenhancements will have significant benefits and equally significant dangers. Some of the benefits will be spectacular in nature, offering wonderful new capabilities and powers. Some of the dangers will be so profound that they will outweigh any benefits, and will therefore justify the banning or postponement of particular categories of modifications.

3. *Its impact will be radical.* Enhancement technologies—even the most apparently sensible and benign ones—will destabilize key aspects of our social order, as well as our understanding of what it means to be human. Many of the values and assumptions that lend structure and meaning to our lives—family bonds, stages of life, our sense of ourselves as mortal and limited creatures, the boundaries between one person and another—will have to be reexamined and perhaps even reconfigured as a result.

WHAT COLOR IS THE TINT OF YOUR GLASSES?

One of the tricky things about looking at the future is that you have to tread a fine line between pessimism and optimism. If you place your accent on the nastier sides of the long human track record—wars, oppression, betrayal, greed, hate—this will inevitably color your image of the future as well. If, instead, you base your vision on the more encouraging aspects of society's past—cooperation, invention, emancipation, mutual understanding—this yields a quite different picture of tomorrow.

In this book I take a middle road, because I believe that the human story is itself—like the enhancement technologies—a mixed bag. History is too complicated (and interesting) to be straightforwardly summed up in a visceral judgment along the lines of "60 percent nasty, 40 percent good," or "70 percent angels, 30 percent devils." Our angels and our devils walk arm in arm all the time, moment by moment.

I have approached this topic not as a historian of science and technology (which is my day job) but more as a middle-aged man who has a daughter and a son, and who found himself fretting about what kind of world they would end up living in. There is plenty for a parent to cringe about in the unknown future, to be sure, but what really bothered me was a nagging thought: what if things actually go pretty well, overall, and our society some-how manages to *avoid* all the more obvious threats that loom over human-kind, such as pandemics, economic meltdown, climate upheaval, large-scale terrorist attacks, or war? What then?

It struck me that, even under this optimistic scenario, we still have some tough challenges coming our way from a rather surprising quarter, namely, medicine and engineering—the arts of healing, building, and inventing. These fields are generally considered among the more benign areas of hu-man endeavor, but they have become so powerful and sophisticated in recent years as to confront us with godlike choices and dilemmas. When do we pull the plug on a stricken relative whose selfhood is no longer intact? What are we to make of a sixty-six-year-old woman who gives birth to triplets? Is it wrong to insert genes from animals and plants into human embryos?

If decisions like these are already baffling and dividing us today, imagine how much more fraught they will be four or five decades hence, when the human applications of biotechnology have come into full deployment. Ma-jor augmentations of our traits and capabilities will spread across the whole panoply of human experience, but will they truly improve our quality of life? When lots of people undertake them—not just hundreds or thousands, but millions—what kind of global social order will result? Might these technolo-gies undermine or disrupt some of the things we hold dearest? Or will they, on the whole, make a positive impact, opening up widespread possibilities for greater human flourishing?

The book has four parts. In the first, "Humans Redesigned," I describe the strategy I have adopted for thinking about the near- and middle-term future. Although none of us has a crystal ball, I argue that not all aspects of the mid-twenty-first century are equally uncertain or unpredictable—and that some forms of disciplined speculation about the shape of tomorrow's society can therefore prove useful. I then survey the state of the art in the science and technology of human bioenhancement, concentrating on three main areas—pharmaceuticals, bioelectronics, and genetics—through which the most far-reaching modifications of humans are likely to occur over the com-ing decades. My emphasis here is on what the experts themselves are saying about the likely future trajectory of innovation in their fields.

In the second part, "Justice," I explore the collective, societal implica-tions of enhancement technologies, while part three, "Identity," focuses on their more ethereal effects on us as individual persons. I avoid sweeping generalizations about the morality of the "enhancement enterprise"—a Manichean thumbs-up or thumbs-down. Instead, I examine each type of enhancement technology on a case-by-case basis. Taken together, these chapters are designed to complement each other like pieces of a jigsaw puzzle, leaving the reader with a vivid sense of what the middle years of the twenty-first century might plausibly look like, and what it would feel like to live in such a world.

Finally in the last part, "Choices," I present an overview of the ethical questions raised by the enterprise of human enhancement. Here I compare the relative merits of radical enhancements as opposed to more modest and limited interventions, posing a set of basic, "civilization-shaping" questions: How far do we want to go in using biotechnology to alter the human constitution? What kinds of modifications should we embrace, and which should we reject? And how much of a say do we really have in determining the manner in which these devices and practices will enter our lives?

In order to lend greater concreteness to the discussion, I open some chapters with brief fictional vignettes in which I describe what the future world of enhanced humans might actually look and feel like, as I did in the opening of this introduction. To be sure, we have no way of knowing which aspects, if any, of these imagined scenarios will actually come to pass. However, as long as we bear in mind that these vignettes are speculative exercises—educated guesses—they can vividly illustrate the *kinds* of challenges that our grandchildren may find themselves encountering.

COMPANION WEBSITE: *WWW.OURGRANDCHILDRENREDESIGNED.ORG*

Since this book explores many domains of cutting-edge science and technology, I have created a companion website where I will post regular updates on the rapidly evolving developments at the forefront of these fields—as well as their social and moral implications. On this site I also present two complete versions of the book's bibliography, the first organized thematically and the second alphabetically, as well as a number of appendices in which I explore in greater depth certain arguments or issues that I excluded from the main body of text for reasons of space. I also include a "dialogue page" where I will respond on an ongoing basis to criticisms, observations, and new ideas provided by readers of the book. You can access the website at www.ourgrandchildrenredesigned.org.

Part I

HUMANS
REDESIGNED

. . . . .

ONE

Envisioning the Future

Between the Jetsons and the Singularity

The future enters into us in order to transform itself in us long before it happens.

—Rainer Maria Rilke, *Letters to a Young Poet*[1]

WHY SOME FORECASTS SUCCEED (WHILE OTHERS FAIL MISERABLY)

Any book about the near- and middle-term future—extending from the present day to several decades over the horizon—justifiably elicits skepticism, because we humans have a very mixed record in the art of prognostication.[2] Some thinkers seem to have foreseen the shape of things to come with uncanny accuracy: one thinks of Alexis de Tocqueville's famous prediction in the 1830s that the rivalry between Russia and America would one day come to dominate world affairs. Tocqueville's prediction took 130 years to come true, but when one pondered his words from the perspective of the 1960s, they seemed downright oracular. Most others, however, have not proved nearly as successful in their forecasts. Amid the myriad examples, you might know of Lord Ernest Rutherford, the leading figure in British physics during the 1930s, who categorically dismissed the possibility of harnessing the forces of the atom as "the merest moonshine."[3] Just twelve years later, this moonshine had obliterated two Japanese cities.

So we face an important question, right at the outset: Is this whole exercise a waste of time? Precisely because our knowledge of the future cannot help but be speculative and imperfect, should we abandon the effort to make out what lies in store for us in the middle years of the coming century? Does that future simply lie too far away for any assessments of it to be of much use at all?

The key to answering this question lies in the fact that not *all* aspects of the future are equally uncertain. Will gravitation still play a role in our lives two hundred years from now? Of course it will. Will someone named Marcia be president of the United States two hundred years from now? No one can know. The reason for the difference is that, in the case of gravitation, we are talking about a *structural* factor: one of those causal processes that only changes very gradually, if at all. In the case of who will occupy the Oval Office, we are talking about a *conjunctural* factor: a causal process that will be determined more locally, by events unfolding in a much shorter span of time.

History, in other words, is usefully thought of as a layered phenomenon, governed by distinct time frames of change: some causal sequences play themselves out over much longer spans of time than others.[4] Therefore, in making statements about the future, it is important to keep the time frame of your prediction commensurate with the causal processes you're describing. If you want to predict whether Marcia will be president, you cannot do so beyond a time frame of one to five years at most. If you want to predict the impact of gravitation, you can safely extend your forecast out for centuries (indeed, millennia) into the future, with confidence that the underlying causal processes will not have had enough time to change in any fundamental way.

Therefore, as we try to envision what the world will look like in the middle years of the twentieth century, we can be reasonably confident in projecting certain structural features rooted in time frames of change of fifty years or more, and proportionately less confident in predicting other structural elements whose qualities are determined in relatively shorter time frames (say, twenty to fifty years). We should avoid altogether the attempt to forecast those conjunctural elements that will be shaped in a causal frame of one to twenty years.

If we divide the events and causal processes of the coming century into these rough categories, what sorts of examples do we get?

LONG STRUCTURAL PROCESSES (FIFTY YEARS OR MORE)

- Physical and material constraints, such as the relation between greenhouse gas emissions and global warming.
- Sociological patterns, such as the division of society into economic classes.
- Deep cultural patterns, such as the division of society into identity groups (for example, religion, race, ethnicity, and so on).
- Gradual economic processes such as globalization.

- Psychological constants in human behavior, such as the tendency to put family first.
- Geopolitical patterns pertaining to the relative weight of particular nations in world affairs.

SHORT STRUCTURAL PROCESSES (TWENTY TO FIFTY YEARS)

- Life span of particular governmental and economic institutions, such as the Environmental Protection Agency (EPA), NASA, the Transportation Security Administration, Pan American World Airways, Starbucks, Microsoft.
- Economic patterns, such as the ascendancy of particular industries or sectors of production (the decline of labor-intensive agriculture, the service economy, the information economy, for example).
- Social and cultural shifts, such as the impact of the sixties counterculture or the changes wrought by the civil rights movement in the lives of African Americans.
- Political patterns, such as the ascendancy of liberal versus conservative ideology in public affairs (Franklin Delano Roosevelt's legacy, the Ronald Reagan era, and so on).

CONJUNCTURAL PROCESSES (ONE TO TWENTY YEARS)

- The influence (for good or ill) of gifted individuals, such as Napoleon Bonaparte, Adolf Hitler, or Mikhail Gorbachev.
- Short-term economic and political patterns, such as economic recessions and recoveries, or the policies of particular presidential administrations.
- Rare but significant events in the material world (earthquakes or disease pandemics, for example).
- Relatively sudden scientific or technological breakthroughs, such as the creation of the atomic bomb, the development of antibiotics, the discovery of DNA, or the emergence of the Internet. We know such short-term breakthroughs will almost certainly happen, but we cannot have a clear idea in advance of what they will be, and how they will affect the future.
- Major political events such as wars, which can change the course of history in unforeseeable but profound ways.

The first feature worth noting about this three-tiered list is that the conjunctural processes are not necessarily superficial in their impacts. While it is true that the vast majority of conjunctural events only exert a relatively local effect on the surrounding world, some of them end up producing major inflections in the course of history. "Conjunctural" does not necessarily mean "ephemeral" or "weak."

This is, of course, bad news for us prognosticators. If the conjunctural processes are the hardest to predict beyond a time horizon of twenty years, and yet some of them prove to be extremely potent determinants of historical change, this introduces a major wild card into all our middle-term predictions. I don't see any way around this basic fact, and it should induce us to be proportionately humble in our efforts of forecasting beyond the twenty-year horizon.

But, here, another aspect of our three-tiered list comes into play. Novel technologies tend to change more quickly, radically, and unpredictably than human social, economic, and cultural institutions.[5] The rise of the Internet, for example, became a major historical phenomenon in less than twenty years, but phenomena like racial prejudice, class conflict, gender bias, and similar social and cultural factors tend to evolve much more slowly. The implication is clear: we are likely to do better at predicting general patterns of human social behavior than detailed technological outcomes. In envisioning the coming century, we should not try to foresee too precisely which technologies will exist in different decades. If we insist on doing so, we are likely to end up pulling a Rutherford.

Instead, we should focus on the impacts that broad categories of technology are likely to have on our society, economy, and culture. We can have no clear knowledge about what the machines of 2040 or 2050 will be like, but we do have a much better sense of how the people of that era would likely respond to specific sets of machine capabilities. We should therefore lay out a deliberately wide range of plausible scenarios for the technologies that might exist in the coming century, so that the very breadth of our range compensates for our inherent weakness in forecasting the fine-grained details of technological change. Then, for each plausible *type* of technology, we should assess what the social impacts are likely to be. This is the procedure I adopt in the chapters that follow.

THE *JETSONS* FALLACY

In 1962 a new TV show entered American living rooms. Modeled on the immensely popular *The Flintstones*, it depicted the daily life and middle-class misadventures of a family of the year 2062, the Jetsons. The show's creators,

Hanna-Barbera Productions, clearly bent over backward to offer the most spectacularly bland image they could of the world that awaits us. A full century will go by, they seemed to be telling us, and absolutely nothing of any importance will change in the slightest. Benignly patriarchal fathers will still bumble their way through the challenges of parenting and office work; wives will be well coiffed; children will cultivate an endearing rebelliousness (within carefully observed limits). What marks the future as different, according to Hanna-Barbera Productions, is the advent of new gadgets. Houses, stores, and office buildings rise up on pillars in the sky; cars fly; children are whisked off to school in pneumatic-tube bubbles. Robots are everywhere, of course. They perform all manner of menial tasks, offering abundant opportunities for hilarity when they malfunction.

This popular TV show perfectly captures a recurring misconception in contemporary visions of the future: the tendency to imagine that *our technologies will evolve dramatically and even radically, while we humans will stay fundamentally the same.* I refer to this as the *Jetsons* fallacy. If one surveys the most influential science-fiction visions of the future—those movies and books that have broken out of the niche of sci-fi buffs and shaped the imagination of the broader society—one encounters this phenomenon again and again.[6]

In the *Star Wars* films and *Star Trek* TV series, for example, we certainly do encounter many alien biological species and intelligent robots.[7] But all these strange beings coexist *alongside* a population of unmodified humans who are no different from the humans of today. Some popular science-fiction movies, such as *Blade Runner*, *AI*, and *Bicentennial Man*, explore the rise of robots or artificial intelligence to a human level of conscious awareness. But here, once again, these advanced machine beings coexist with a population of humans who remain rigorously unmodified, undefiled.

In many cases, as in *The Six Million Dollar Man*, *The Bionic Woman*, *Hulk*, *Inspector Gadget*, *Alien: Resurrection*, *Spider-Man*, *Iron Man*, or *Limitless*, we do encounter humans who undergo profound biological redesign, but always under the same conditions: whether accidental or deliberate, the modification is uniquely confined to a single individual, and never applies to the surrounding population. Viewers are emphatically *not* invited to contemplate a world in which large numbers of humans have gone down this path.

Aldous Huxley's 1932 masterpiece, *Brave New World*, stands out as the most important exception to this trend. Huxley's imagery of bioengineered castes and drug-induced manipulation of the masses has penetrated very deeply into contemporary culture. What is striking, however, is precisely how unique *Brave New World* remains within the subset of science-fiction works that have become cultural landmarks. Eight decades have passed since its publication, and it still occupies a place of solitary prominence as the only systematic effort to imagine a world populated by engineered humans.[8]

One might object that the 1997 film *Gattaca* fills such a role. But this film, too, falls into its own version of the *Jetsons* fallacy. In the world it depicts, genetically enhanced humans and unmodified humans are impossible to tell apart (except through DNA tests). The movie tells the story of an unmodified human who, through sheer force of will, overcomes his genetic "destiny" and outperforms those engineered humans whom society labels as superior. This is certainly a dystopia, but it is a dystopia that refuses to confront the consequences of its own underlying premise: the possibility that genetic science, systematically applied, might actually result in a population of individuals whose mental and physical performance vastly exceeds that of the unmodified. In this sense, the film's message seems to be that, even if genetic engineering does become widespread, this won't be such a big deal. Ordinary old-fashioned humans will still be able to prevail.

What are we to make of this phenomenon? The answer lies in the economics of mass-appeal films and books, which are supposed to beguile us, not freak us out. The technological modification of entire populations of humans is not a fit subject for mass entertainment because it is simply too disturbing to contemplate. If it must be confronted at all, it must always be coupled with the ironclad assurance that, even in the most extreme future, ordinary humans will still exist, and will still continue to be what they are today.

The only problem with this comforting picture of the future is that it is probably not true! We are headed into a social order whose most salient new feature may well be the systematic modification of human bodies and minds through increasingly powerful means. This process is already underway today and seems unlikely to slow down in the decades to come. The prevalence of the *Jetsons* fallacy suggests that many people in contemporary society are living in a state of denial, psychologically unprepared for what is actually far more likely to be coming their way.

THE SINGULARITY

One writer who has definitely *not* succumbed to the *Jetsons* fallacy is the inventor and businessman Ray Kurzweil. In a series of books written over the past twenty-five years, he has laid out a vision of the future that not only prophesies radical change, but enthusiastically embraces it.[9] Kurzweil conceives of this coming transformation as a relatively sudden event, like the crossing of a threshold: he calls it the Singularity.[10]

The image of the Singularity is drawn from the physics of black holes, which astronomers sometimes describe as "singular" objects in space-time because they are locations in the universe at which the fundamental laws of physics break down. The metaphor is apt, because if people like Kurzweil prove right in their predictions, then we are truly approaching a historical "event horizon," beyond which all extrapolations become meaningless.

On the other side of that divide, Kurzweil believes, the combined impact of genetics, nanotechnology, and informatics will bring about the birth of a radically transformed human species. Humans will redesign their own bodies and minds, using the powerful tools of genetics and nanotechnology; they will reverse-engineer the human brain, applying this knowledge to design new forms of artificial intelligence far more potent than any human mind; they will endow these superintelligent machines with bodies more capable and versatile than any mere biological being could ever hope to be. In this way, humankind will in effect be giving birth to its own successor species, our own technological progeny, whose limitless potential will take them out into the cosmos to fulfill a destiny greater than any of us today can fully comprehend. It will be a moment of *species metamorphosis*, a collective transformation akin to the transition from a caterpillar to a butterfly.[11]

Kurzweil is by no means alone in holding this view. His perspective is shared (to varying degrees) by the proponents of a loose cultural and philosophical current that has come to be known as transhumanism. The transhumanists have formed an organization to promote their ideas—Humanity Plus (or H+)—and though it currently has only some six thousand members worldwide, its influence has been growing rapidly.[12] Many books and articles have been written over the past decade by philosophers, economists, technologists, journalists, or scientists who are engaging in one way or another with these kinds of ideas.[13] I return to the transhumanist philosophy later, but I want to focus here on Kurzweil's writings, for he is without a doubt the most influential exponent of this techno-enthusiastic current of thought about the future.

Kurzweil's argument hinges on a single proposition: historical change and, particularly, technological change are dramatically speeding up, and the *rate* of acceleration is itself increasing exponentially as time goes by. Kurzweil calls this "the law of accelerating returns," and in his book *The Singularity*

Is Near, he provides an abundance of intriguing graphs and charts to make his case.[14] But Kurzweil does not confine himself to modern technological progress: he also extends his "law of acceleration" to a much broader, more metaphysical scale. The history of life, he contends, consists of six major evolutionary epochs, each building on the ones before it and unfolding much faster than the last—from the elemental foundations of physical and chemical processes, to the advent of biological life, to the development of advanced neuronal structures like the human brain, to the rise of modern informatic technology. Extrapolating from these trends into the middle-term future, Kurzweil believes the next steps will involve the "merger of technology and human intelligence," followed by the colonization of outer space by superintelligent beings that humans will soon be creating.[15]

I believe Kurzweil's understanding of historical process and, therefore, his predictive model are deeply flawed. Nevertheless, clarifying the strengths and weaknesses in his argument can help us develop a more refined approach to thinking about the future of technology. Here, then, are three key problem areas in Kurzweil's vision.[16]

First, history is jerkier, more discontinuous, than Kurzweil thinks. Kurzweil's mistake lies in applying his model of accelerating change to historical process as if the passing of the years were a smooth, uniform, and linear phenomenon. But history does not work this way. Its myriad causal processes advance unevenly, with some of them surging steadily forward, others lurching spastically in fits and starts, still others stagnating or regressing—all at the same time.

He seems to conceptualize the major advances of the 1800s and 1900s as if they were stepping-stones in a great march of progress, with steadily increasing densities of change as the decades went by. According to this model, the technological stepping-stones in the 1970s should, in general, be much closer together than the stepping-stones in the 1940s. But this is demonstrably not true. In nuclear physics, for example, the advances of the 1940s were far more rapid and dramatic than those of the 1970s. Kurzweil's concept of "acceleration" is therefore misleading, because it applies as a blanket generalization to all areas of historical change.

Second, history is messier, more interconnected and tangled, than Kurzweil thinks. Kurzweil places considerable weight on the synergies of technological change, as different domains of progress assist and boost each other, further fueling the advance of innovation. Unfortunately, he fails to take into account the obverse of this factor: many historical causal processes do not reinforce each other at all, but rather undermine, degrade, or even nullify each other. The technological writer Bob Seidensticker cites the example of aircraft technology to illustrate this point.[17] Starting with the Wright brothers in 1903, airplanes went through five decades of exhilarating development, their rapidly evolving designs allowing ever greater distances to be traversed at ever higher speeds. If, in the year 1960, one had plotted these innovations on a graph and then projected that trend line into the future, one might reasonably have concluded that supersonic and perhaps even hypersonic aircraft would be the norm by the year 2000. But that is not what happened. Even though the scientific and engineering capabilities for supersonic travel were solidly in place, society made a different choice. Instead of opting for "ever faster, ever higher," it went for "gazillions of passengers, on the cheap." As a result, our plot line of rising aircraft performance dramatically levels off in the 1960s, and remains more or less flat right through the next fifty years.

This is a very different sort of reality than the one depicted by Kurzweil in his graphs and projections. It is a world in which powerful social and economic factors—the oil crisis of the 1970s, the advent of mass air travel, the fickle tastes of consumers—all conspired to lead a particular technology down an unexpected path.

Third, technological advance is not inevitable. In Kurzweil's model, the evolution of devices and machines is driven by two primary factors: the incessant demands of consumers for "more, better, faster," and the ingenuity of scientists and technologists who scramble furiously to meet those demands. Taken together, these forces are so powerful that they render the advance of technological capabilities virtually unstoppable. Resistance is futile.

Most historians of technology, however, do not regard technological change in this manner—as an irresistible tide surging through history, engulfing all those who get in its way. Instead, they see technology and human choices constantly interacting with each other, shaping each other over time.[18] No device exists in isolation from the social and cultural milieu in which it is conceived, designed, manufactured, and used. A hammer, a computer, a robot, an MRI machine—each of these exists as part of a broader system of functions, and those functions are all, without exception, socially determined.

In this picture, human choices play a much larger role than in Kurzweil's. As in the case of the supersonic Concorde, it is quite possible for a majority of rational individuals in a democratic society simply to say no to one kind of progress in favor of another: "more" is not always better, and the most sophisticated technologies do not automatically and unavoidably triumph over the humbler ones. We therefore need to regard Kurzweil's dazzling array of graphs and projections with a healthy dose of skepticism. There is nothing inevitable about our technological future.

For all these reasons, therefore, Kurzweil's vision of an inexorably approaching Singularity is probably inaccurate. This does not mean, however, that his many conjectures about the future are worthless. On the contrary, he and his fellow transhumanists are very much on to something. They have made the empirical observation that science and technology are being turned with growing potency upon humans themselves; and they have concluded that this trend will raise increasingly pressing questions about who we are and what we may become. Both of these are reasonable assessments of where humankind now finds itself.

Our society has gradually put in place a complex and well-funded network of institutions specifically designed to generate rapid innovation in science and technology. In 1850, only 0.03 percent of the total US workforce was employed in fields of science, engineering, and technology; by 1950, this figure had multiplied thirty-six-fold, to 1.1 percent of the workforce; and by 2001, it had grown still further, to 4.2 percent.[19] It is just as serious a mistake to underestimate the significance of this development as it is to exaggerate it. No civilization has ever devoted such vast human and financial resources to the accelerating development of technoscientific prowess.

In the domain of pharmaceuticals, we can now control which part of a chemical substance we wish to activate, and fine-tune the interaction between its molecules and our own cellular processes. Our bioelectronic prostheses are no longer like eyeglasses, adding an external layer to our senses: they now reach deep into our nervous system, changing how it performs in its fundamental workings. Today's genetic interventions penetrate straight to the core, using detailed genomic maps to target specific sites in DNA and redirect their functions along the lines we desire. All these technologies have one key feature in common: they illustrate the *directness* with which humans can now manipulate the innermost workings of nature's processes.

Even though these biotechnological innovations are not advancing smoothly, exponentially, or unstoppably, they are nonetheless harbingers of something big. Kurzweil and the transhumanists are right to get excited about them. This transformation will probably arrive piecemeal, in untidy increments and jumps, extending over a period of many decades through the middle of the twenty-first century. Our children and grandchildren will experience it directly.

TWO

Pharmaceuticals

O true apothecary! Thy drugs are quick.

—William Shakespeare,
Romeo and Juliet, Act V[1]

The concept of "human bioenhancement" is a slippery customer. Its purview ranges from minute adjustments in a person's biochemistry ("I boosted my immune system by sucking on a zinc lozenge") to wholesale redesign ("My latest enhancement package includes infrared vision, Google accessible by thought, and a 160-year health span"). It can refer to modifications of existing traits ("My memory has been far better since the doctors tweaked my hippocampus") or to new capabilities that no human has ever possessed ("This gene retrofit for skin chlorophyll lets me absorb solar energy by photosynthesis"). In this book I adopt the following broad definition: *a modification of a person's biologically grounded traits, adding capabilities that would not otherwise have been expected to characterize that person.*[2]

The three main categories of human enhancement—drugs, bioelectronics, and genetics—bear down upon us along different horizons of time. Major genetic interventions reside for the most part in the realm of future possibilities—perhaps twenty years away, perhaps even more—because the science undergirding them still remains relatively new and uncharted. Bioelectronics is keyed to a nearer future: we already possess functional devices today, such as brain-machine interfaces for the paralyzed and the blind, that presage significant breakthroughs in the conjoined fields of neuroscience and informatics within the next ten to twenty years.

Pharmaceutical enhancements, by contrast, are already very much part of our contemporary world—and have been for quite some time. I remember in 1966 hearing the new Rolling Stones song "Mother's Little Helper" on the radio. Even as an eleven-year-old, I understood the song's message: Mother wasn't actually sick, but a little yellow pill would propel her through her hectic day. The Stones' sardonic lyrics illustrate the extent to which a modest form of pharmacological enhancement had already become a widespread phenomenon by the 1960s.

In this chapter I briefly sketch the trajectory of our society's use of such chemicals over the past century or so, focusing primarily on their use for enhancement purposes rather than therapeutic treatment (though the line between the two is often blurred in practice). Then, having completed this survey of past and present, I turn toward the prospects for chemical enhancement over the decades to come.

A HUNDRED YEARS OF PILL POPPING

Virtually all the statistics on pharmaceutical usage during the twentieth century indicate continual and dramatic growth. The variety of drugs available; quantities of pills taken per capita; number of individuals relying on medication; dollars spent on research, manufacturing, and advertising; as well as the level of cultural attention paid to chemical intervention—in all these areas the twentieth century can be described as a period of explosive expansion.[3] The drugs themselves, moreover, have been getting ever more effective as time has gone by, becoming increasingly precise in achieving their stated purpose while minimizing unwanted side effects.[4]

A key phenomenon here is that of "medicalization"—the growing use of drugs for relatively common conditions such as lethargy, distractedness, or moroseness that would once have been considered mere character flaws.[5] If you are very shy, for example, you can go to a psychiatrist, who may diagnose you with "social anxiety disorder." You will then get a prescription for a drug like Nardil, which stands a good probability of making your shyness melt away like ice in the sun.[6] A condition that would once have been regarded as a rather challenging facet of one's personality, to be managed through self-discipline and grit, has now become a diagnosable disorder for which powerful medications can be prescribed.[7] Pill popped, problem solved.

Taking drugs to tweak your physical or mental performance no longer bears the social stigma it once did. The scholar John Hoberman argues that a culture of "lifestyle medicine" has emerged over recent decades, a moral milieu in which individuals feel empowered to reengineer key aspects of their selfhood in line with idealized notions of their own potential. This trend applies not just to pharmaceutical enhancements such as steroids or hormones, but is also evident in the dramatic rise of cosmetic surgery, diet aids, body-building devices, sexual augmentation, and antiaging remedies that have become common aspects of today's world.[8] Many of these practices are now widely regarded as perfectly legitimate and reasonable avenues of self-improvement and self-fulfillment.[9]

Enhancement drugs can be divided into four broad groups, depending on whether the traits they modify are primarily physical, cognitive, or emotional in nature, or whether their target is the more global phenomenon of aging itself.

Physical traits

On September 21, 1998, the newspaper columnist Ann Landers printed a letter she had received from a group of elderly women in California about the newly released medication Viagra. The women, who called themselves the Senior Señoras in Sonoma, took a straw poll among themselves and reported the results: three of them stated that they enjoyed sex and welcomed their husbands' use of the drug; eight of them begged to differ and hoped their husbands would not start taking it; another four went even further, making it clear that they had "endured" sex all their lives out of a sense of duty, but now wanted once and for all to be left in peace. They concluded their letter with the observation that "Viagra must have been invented by a man."[10]

No enhancement, it turns out, is ever purely physical in nature: to paraphrase Freud, an erection is inevitably more than just an erection. Humans are cultural creatures, and all aspects of our bodily life are filtered through the sentiments and meanings with which we cannot help but imbue them. Hence, to the Senior Señoras, Viagra represented something quite different than it did to their menfolk: for most of these women, who believed in a certain "natural" progression of life stages, the drug became a royal pain in the neck, forcing them to renegotiate aspects of their identity that they had considered stable and resolved.[11] All this, in a little blue pill.

The number and variety of such pharmaceuticals have been growing rapidly over recent decades. In the category of cosmetic enhancements, we find drugs for body building (anabolic steroids), shortness (human growth hormone), slimming (Xenical, Meridia, Belviq, and Qsymia), baldness (Propecia), and the elimination of wrinkles ("gene foods").[12] In the category of functional enhancements, we find drugs for athletic performance (erythropoietin, human growth hormone, androstenedione), heightened energy (amphetamines and their many derivatives), hormone replacement (estrogen, testosterone), and augmented sexual vigor (Viagra, Cialis, Levitra).

A skeptic might take the position that there is nothing really new going on here. As far back as the historical record can show, we find evidence of people in past civilizations using stimulants such as tea, coffee, tobacco, and coca to boost their energy levels, as well as other tisanes and dietary supplements to augment (or at least attempt to augment) various aspects of their physical performance. What is different today, however, is the potency with which these self-modifications can take place. Propecia, unlike the various snake oils of centuries past, really does compel the body to grow new hair. Erythropoietin, unlike the unguents and herbal infusions adopted by athletes in the past, triggers a major change in blood chemistry: the sudden profusion of invigorating red cells can render the blood so thick that it actually endangers the athlete's life. Viagra incurred the ire of the Senior Señoras precisely because it worked extremely well.

The result, not surprisingly, has been a new level of social and cultural upheaval. In the world of competitive sports, for example, it is hard to find a serious commentator who does not believe the situation has reached a crisis point. World-class athletes forced to retire in disgrace; demoralization among the competitors; an ongoing arms race between regulatory bodies and the developers of illicit pharmaceuticals, as they struggle to stay ahead of each other—and no end in sight.[13] A cloud of suspicion and confusion has fallen over the entire enterprise of competitive athletics, as both spectators and competitors are forced to confront thorny philosophical questions: What is sports really about? Is it about performance? About teamwork and self-discipline? About the cultivation of innate talent, or the bottom line of winning? One finds no easy answers here. What *is* clear is that these pharmaceuticals have already destabilized the moral frameworks surrounding one of the most ancient and beloved arenas of human social life.

Cognitive traits

"Viagra for the Brain," run the headlines. "Smart pill." "Neuroceuticals." It is hard not to share the sense of giddy excitement that has surrounded the fields of augmented cognition, learning, and memory in recent years.[14]

1. *1990*: the neuroscientist Eric Kandel identifies a brain chemical known as CREB as a key factor in the processes of memory formation.[15]
2. *1994*: the neuroscientists Tim Tully and Jerry Yin alter the way CREB operates in a strain of fruit flies. The CREB-boosted flies learn new

behaviors at rates up to ten times faster than their unmodified brethren. Other researchers later replicate these findings in mice.

3. *1999*: Princeton neuroscientist Joe Tsien genetically engineers a strain of mice possessing extra copies of the NR2B gene, which regulates the activity of synaptic receptors in the brain. The NR2B-enhanced mice perform up to five times better than unmodified mice in tests of memory and learning.

4. *2000s*: Several new drug companies race to apply these findings to humans. Memory Pharmaceuticals (founded by Kandel) and Helicon Therapeutics (founded by Tully) both have CREB-related medications going through clinical trials. Another company, Cortex Pharmaceuticals, is exploring a brain chemical known as ampakines, aimed at producing similar results through a different neural pathway.[16]

The notion of boosting memory in humans sounds at first like a terrific development. I would be able to learn foreign languages faster, recall more accurately the names of people and places I have known, find my car keys without a lot of cursing and fuss. But what about forgetting? When we examine the functioning of memory as a practical component in a person's daily life, we find that it is just as important to be able to selectively *lose* information as it is to retain it. Without this ability, we would rapidly find ourselves drowning in a sea of trivial details, impressions, emotions, and images. When researchers seek to enhance the functioning of memory in humans, therefore, they will actually be required to do something far more complex than merely augmenting carrying capacity—as if they were doubling the size of a hard drive in a computer. Rather, they will need to boost carrying capacity, search-and-retrieval function, the ability to triage items according to importance, and the deletion function all at the same time—for this nexus of interconnected abilities is what we really mean by the word "memory" in the context of human awareness.[17] This is obviously a much taller order than it at first appears.

Already, however, a first generation of "smart pills" beckons those citizens who would like to modify their cognitive profile.[18] None of these chemicals was originally intended as an enhancement pharmaceutical: they were developed to help people with basic functional deficits, such as narcolepsy, extreme restlessness, or an inability to stay focused. Nevertheless, for many people who have no such deficits, they also seem to work remarkably well in boosting cognitive ability.

The most commonly used pharmaceuticals in this category are the stim-ulants Ritalin (methylphenidate) and Adderall (dextroamphetamine), which are widely prescribed for patients with attention deficit hyperactivity disor-der (ADHD). Ritalin has been in use since the 1960s, whereas Adderall came onto the market in 1996; both have relatively few side effects, though they are potentially habit forming if used over long periods. A related stimulant medication, Provigil (modafinil), developed in 1990, has met with consider-able success in the treatment of narcolepsy and other sleep disorders.

When healthy people take one of these drugs, they tend to report two quite different kinds of results. Some find that the chemicals do little more than render them alert and focused over longer periods of time.[19] But others encounter a quite different experience. Björn Stenger, a doctoral candidate in computer engineering at Cambridge University, took a single dose of modafinil as part of a research study. Within one hour, he observed, his at-tention and memory both seemed noticeably sharper. He then went on to perform at very high levels on a series of mental agility tests. Overall, he said, "if [my] brainpower would normally rate a ten, the drug raised it to fifteen. It was kind of fun."[20] Another person, who responded anonymously to a 2008 questionnaire on the website of the *Chronicle of Higher Education*, affirmed that going on Adderall had changed his career: "I'm not talking about being able to work longer hours without sleep (although that helps). I'm talking about being able to take on twice the responsibility, work twice as fast, write more effectively, manage better, be more attentive, devise bet-ter and more creative strategies."[21] Here, in other words, it is the higher cognitive functions such as learning, memory, and insight that are purport-edly being affected by these enhancement drugs.

Scientific assessments of these chemicals' effectiveness have yielded simi-larly contradictory results. One study conducted in 2003 at the Walter Reed Army Institute of Research concluded that the effects of modafinil on a per-son's ability to function while sleep deprived were about the same as a high dose of caffeine.[22] Another series of experiments, conducted at Cambridge University in 2001, reached a very different conclusion: after taking 200 mil-ligrams of modafinil, test subjects "moved more smoothly from one task to the next and adjusted their strategies . . . with greater agility. In short, they worked smarter and were better at multi-tasking."[23]

Given these conflicting conclusions, it would make good sense to with-hold judgment about cognitive enhancers until the results of further re-search are in. But "wait and see" is clearly not what a great many people in today's society are choosing to do: they are scrambling to get these drugs in any way they can, and the rates of illicit or off-label use in the United States over the past decade have been skyrocketing. Estimates vary: one 2011 survey concluded that on some campuses, up to 30 percent of students have been using cognitive enhancers on a fairly regular basis.[24]

Emotional traits

In the summer of 1996 the psychologist Jonathan Haidt did a little experiment on himself: he went on Paxil, a mood-altering drug similar to Prozac. He was not suffering from depression or any other debilitating condition, but he was curious to see what all the fuss over these kinds of drugs was about:

> It was like magic. A set of changes I had wanted to make in myself for years—loosening up, lightening up, accepting my mistakes without dwelling on them—happened overnight. However, Paxil had one devastating side effect for me: it made it hard for me to recall facts and names, even those I knew well. . . . I decided that as a professor I needed my memory more than I needed peace of mind, so I stopped taking Paxil. Five weeks later, my memory came back, along with my worries. What remained was a firsthand experience of wearing rose-colored glasses, of seeing the world with new eyes.[25]

Haidt had been doing just fine in his life, but he wanted to see what it would mean to do "better than fine." His story vividly raises the core issue surrounding the enhancement of human affect through pharmaceuticals: the question of authenticity. Which fellow is the real me? If Haidt had not encountered the side effect of poor memory and had therefore stayed on the drug (as he unabashedly affirmed he would have done), which man would have been the more authentic one: the anxious fellow coping as best he could with his tenure worries, or the cheery young professor heartily enjoying the early years of his career? Is there something inherently phony or disreputable about a state of lightheartedness attained through chemicals?

The bioethicist Leon Kass argues that the answer is definitely yes: "We want to perform better in the activities of life—but not by becoming mere creatures of our chemists or by turning ourselves into bionic tools designed to win and achieve in inhuman ways. . . . We want to be happy—but not by means of a drug that gives us happy feelings without the real loves, attachments, and achievements that are essential for true human flourishing."[26]

The image Kass seems to have in mind here is Huxley's *Brave New World*, with its masses placated by the drug soma, their lives repulsive to us in their drudgery and meaninglessness.[27] The people in Huxley's dystopia declare themselves to be subjectively fulfilled, but their lives give no grounds for such fulfillment. Their loves and friendships are superficial, their thoughts and impulses are channeled by conditioning, and their work is routinized and unchallenging. It is this deep alienation that leads Kass, with justification, to hold this up as a canonical exemplar of inauthentic existence.

But here's the rub: Huxley's dystopia is, by design, an extreme portrayal. Why should we believe that *all* chemical interventions for mood control necessarily lead down a path to the Brave New World? Haidt's experiment with Paxil suggests a very different set of questions. What if the drugs, in some cases, end up rendering the life of an already healthy person even more meaningful than before? What if they enable people to have more enduring and engaging relationships, better and more challenging work, more imaginative and satisfying play? What if the chemicals allow us to gain a greater measure of control over destructive psychological patterns that have defeated all our past efforts to reorient them? I'll return to these questions later.

Pharmaceuticals against aging

Getting old, as anyone my age knows, is the ultimate un-enhancement. All traits—physical, cognitive, emotional—come under assault. What if we could dramatically increase not just our life span but our *health span*—the number of years we live in full possession of our physical and mental faculties?

Gerontologists and other biological researchers have made great strides in the past two decades toward understanding the phenomenon of aging, though they acknowledge that many basic questions still remain unanswered.[28] Two models of human aging have emerged—the "wear and tear" theory and the "programmed senescence" theory. Although these approaches are distinct, most researchers believe that both kinds of factors probably play key roles in taking us toward decrepitude.

Starting even before the day we are born, the human organism is subjected to a wide variety of what biologists call "insults"—stresses and damages that undermine the functioning of our cells.[29] For the most part, the human organism succeeds quite well at keeping these wear-and-tear processes at bay until we reach our midtwenties or early thirties. Defects are repaired; toxins are neutralized; malfunctions are reversed. Then, mysteriously, the fix-it mechanisms begin to falter, and the insults commence taking their toll. This deterioration occurs at different rates in different people, but sooner or later it affects us all, with ultimately fatal consequences.

Why? What evolutionary advantage could there possibly be to have us winding down like spring-powered clocks after a certain age? The answer is straightforward: evolution did not design us to get old, but rather to stay healthy until we could reproduce.[30] After we reach reproductive age and stay there a few years, evolution does not care a fig what happens to us. All the selective pressures that operate to keep our organism fit and resilient dwindle away after we reach our midtwenties.

How, at that specific point in life's trajectory, does the body come to stop repairing itself, allowing its various systems to commence deteriorating? Here we encounter the second model of aging, the theory of programmed senescence. It has two main elements.[31]

- *Telomeres*. At the end of every human chromosome lies the same repeating sequence of nucleotides: TTAGGG. This chunk of six genetic letters repeats about 1,500 times, comprising what is known as the telomere, the end-cap at the chromosome's tip. Every time the cell divides, the telomere loses exactly one TTAGGG segment: this progressive shortening of the telomere operates as a cellular countdown mechanism. After a specific number of cycles of cell division (reflected in a proportionately shortened telomere), the cell suddenly stops dividing. It does not immediately die, but it enters a state of senescence and never reproduces itself again. The cell has reached its Hayflick limit (named after the biologist Leonard Hayflick, who first described the phenomenon in 1965).[32] Scientists believe this telomeric countdown, continually operating among most of the sixty trillion cells that make up our body, plays a key role in the aging of the whole organism.
- *Genetic triggers*. Part of the normal development of a human body involves a phenomenon known as apoptosis, or programmed cell death. When a cell begins malfunctioning, or when its existence is no longer needed, it is targeted by genetic messengers that compel the cell to cease functioning and die.

Experiments with both these types of cellular-programming mechanisms have yielded exciting results. In 1998, scientists at Geron Corporation inserted a telomere-repairing enzyme known as telomerase into healthy human cells and then let the cells go through their normal division process. To their delight, they found that these telomerase-enhanced cells blew right past the Hayflick limit and continued vigorously splitting and replicating, time after time, as if all constraints on their life span had evaporated.[33]

Another series of landmark experiments involved manipulations of DNA in a small roundworm, or nematode, known as *Caenorhabditis elegans*. One such experiment, knocking out a gene known as daf-2, resulted in nematodes that lived twice as long as their unmodified compatriots. Another experiment, focused on the insulin-regulation gene age-1, resulted in worms that not only lived longer but also showed heightened resistance to external stressors. Steven Austad, a biologist at the University of Texas, articulated the sense of awe that many scientists felt as they encountered these results: "Worms have 100 million nucleotides in their genome approximately. If you change one of those, one of the 100 million letters of the genetic alphabet, you get a doubling of lifespan. If you change two of the letters, if you change one letter in one gene, one letter in another gene, you'll get a tripling of lifespan."[34]

Over the past fifteen years, similar genetic interventions have resulted in significantly extended life spans for yeast cells, fruit flies, and mice as well.[35] But will it work in humans? The notion of dramatically altering the human health span, not just by a few years but by doubling or tripling it, remains controversial today. Some, like Hayflick, continue to believe after decades of research that this idea is nothing but science fiction.[36] Others take the very different view articulated by Michael Rose, a biologist at the University of California, Irvine: "In twenty-five years we could see the creation of the first products that can postpone human aging significantly. This would be only the beginning of a long process of technological development in which human lifespan would be aggressively extended. The only practical limit to human lifespan is the limit of human technology."[37]

What is striking here is not so much the ongoing controversy over this issue, but the fact that so many scientists are now heatedly debating it at all. Three decades ago, the vast majority of biologists firmly believed that the human life span, like that of most other species, fell within a specific and ultimately unalterable range. Now that certainty has crumbled. The possibility of significant life extension lies squarely on the agenda of contemporary biological research, and dozens of labs all over the world are working on various aspects of it. It is now widely regarded as a fascinating hypothesis to be taken seriously and explored.[38]

WHERE WE MAY BE AT MIDCENTURY

While it would be foolhardy to attempt any detailed predictions for the coming century, it seems reasonable to expect that all four of the thematic areas I've surveyed will provide ever greater opportunities for chemical augmentations of human performance. Three factors will render this possible: our ever-deepening understanding of human physiology and brain function; our clearer grasp of how each physical and mental capability correlates with specific biochemical processes; and our increasing ability to intervene, with chemicals targeted at the molecular level, to modulate those underlying processes with great precision. This has been the unmistakable trajectory of the past half century—better knowledge, sharper tools—and there is no reason to believe the trend over the following decades should prove any different.

Our physical traits will be open to more direct and potent sculpting than ever, probably with fewer side effects than today. Cosmetic enhancements will proliferate, allowing us not only to choose our appearance, but perhaps even to modify it many times over. Functional enhancements will allow us to override some of the limitations that biology and luck impose on today's humans. If we so desire, we will be stronger, faster, and, yes, erecter. The world of competitive athletics will have to adapt: it may even have to evolve into a two-tiered system, with separate sports events for the enhanced and the unmodified.

The manipulation of human cognitive abilities will spread beyond college campuses and become part of everyday life for many people, whether they like it or not. Chances are, however, that a majority of individuals *will* like it very much indeed and will come to consider the chemical boosting of mental performance perfectly normal and acceptable. Attention, learning, memory, or insight—each of these cognitive faculties may have to be augmented in different ways, through distinct chemical pathways. Perhaps, however, universal cognitive enhancers will also have been developed, offering a more generalized heightening of mental functions.

By midcentury, or thereabouts, we may have the ability to fine-tune our own moods and feelings with unprecedented precision. As neuroscience and cognitive psychology come to offer an increasingly sophisticated understanding of the functional architecture of the human brain, this will translate into new methods for modulating the brain states and hormonal functioning that underpin emotion.[39] Mood-altering drugs are likely to become far more effective than they are today—both for heightening the vividness and nuance of our affective experiences, and for damping them down when we so desire.

Aging will still occur, but we will probably be offered new products that dramatically slow down the deterioration and attenuate its effects. The average life span and health span of the population will continue to grow. Some aspects of our physical and mental profiles may even prove amenable to moderate forms of systematic rejuvenation. In all likelihood, the more radical dream of the visionaries—tripling the human health span or abolishing aging altogether—will continue to elude us in the second half of the twenty-first century. But this is far from certain: it remains one of the more significant wild cards in our future.

THREE

Bioelectronics

I'm not *me* anymore. I'm a hardware store!

—Inspector Gadget (Matthew Broderick)
in *Inspector Gadget* (1999)[1]

The research in bioelectronics being done today in labs around the world veers sharply into areas long associated with science fiction. If we go on YouTube, for example, we find a 2012 video of a woman controlling a robotic arm purely by thought.[2] Her name is Cathy Hutchinson, and she's been paralyzed from the neck down since suffering a massive stroke in 1997. Brown University neuroscientist John Donoghue has hooked her up to an advanced brain-machine interface. Sensors implanted in her motor cortex are picking up her brain activity and then sending the signal down wires to a computer, which decodes it and transmits it to a free-standing robotic arm across the table from her. Cathy frowns with concentration, and the arm glides down, extends its humanlike hand toward a metal bottle on the table; the fingers gently close around the bottle's sides; then slowly, smoothly, the hand travels across from the table toward Cathy, proffering her the bottle. There's a plastic straw sticking out of it. She tilts her head, brings the bottle a bit closer, closes her lips around the straw, and takes a sip. The researchers all around are watching in silence, mesmerized. She lets the straw go, the robotic hand glides back to the table, sets the bottle there, and lets go. Cathy looks over at the camera and breaks into a broad grin. The room erupts in applause. It's the first time she's been able to give herself a drink in fifteen years.

It is hard not to view such videos without reaching an entirely sensible conclusion: what was nothing more than a sci-fi special effect in the past—the control of machines by thought—is now becoming a reality. Then comes the next logical step. If scientists can tap directly into someone's brain, allowing her to maneuver a robotic device, what other machines might people learn to control by thought alone? How long will it be before you or I can control a car, a mechanical exoskeleton, an android robot, an unmanned drone flying in the sky above?

And then the next logical step after that: if we can read the internal signals the mind uses to control motion, could we not also (perhaps) read the different sets of signals it uses for purposes of communication? Wired and connected to each other through a computer interface, could you and I directly share our thoughts, memories, or feelings, brain to brain?[3]

The scientists themselves are the first to warn us about leaping to these kinds of conclusions. No, no, no, they tell us. We are not trying to build superhuman capabilities, but merely to restore lost function to people who have suffered disabling accidents or diseases. And we are only at the beginning, they insist. We still have so much to learn, so many extraordinarily difficult problems to solve. But then one looks again at the videos on YouTube, and one cannot help but reflect: look at the problems these scientists have *already* solved.

BIOELECTRONICS: TODAY'S STATE OF THE ART

My definition of "bioelectronics" encompasses all applications in which electrical, prosthetic, or informatic devices connect directly with the human brain or sensorium to achieve new functional effects.[4] Such devices are currently being developed for a wide variety of purposes: boosting the senses, controlling machines, augmenting memory and cognition, and monitoring or even manipulating the internal states of the mind.

Restoring the senses

Jens Naumann, a thirty-nine-year-old Canadian farmer, lost his sight in two separate accidents between the ages of seventeen and twenty. He was completely blind for twenty years. Then, in 2002, he volunteered to have the world's most advanced visual prosthesis—built by a researcher named William Dobelle—implanted inside his brain.[5] The Dobelle system required Naumann to wear a special pair of glasses on which a small video camera had been mounted.[6] The camera's video signal passed through wires to a computer worn on the hip, where it was translated in real time into a stream of electrical pulses. From there, another series of wires carried the pulses up, through a hole in Naumann's skull, into an electrode array resting directly on the surface of his visual cortex at the back of his brain. Sixty-four electrodes delivered carefully calibrated micropulses of electricity to the visually tuned neurons of the cortex.

The day the system was activated, Naumann went outside the lab into a nearby parking lot, accompanied by Dobelle and a reporter. Someone handed him the keys to a blue Mustang convertible. He got into the driver's seat and turned the key. One of Dobelle's assistants filmed Naumann with a video camera as he slowly pulled away.

You can watch him online, in a clip lifted from CNN.[7] A blind man driving a car. He sits, tautly leaning forward, scanning left and right. The vehicle is going about five miles per hour. He approaches the end of the lot, makes a smooth left turn, staying in the lane, and heads down the clear space there. Then he stops and comes back in reverse, leaning out the window and peering behind the vehicle, his black glasses protruding like goggles from his face.

Today, more than a dozen labs around the world are feverishly working toward the same goal as Dobelle, exploring a wide variety of techniques and methods.[8] Some, like the Utah visual neuroprosthesis program, are working to refine Dobelle's idea of an implantable electrode array.[9] Others have adopted an alternative approach, hypothesizing that it might be better to insert the implant directly within the damaged eye and then let the eye's natural connection to the rest of the brain's visual system carry out the signal processing.[10]

Controlling machines

The big problem with brain implants like those of Hutchinson and Naumann, however, is that they fall under the category of "invasive technologies": they require difficult, messy, and expensive surgical procedures that inflict severe trauma on the body. Moreover, the implanted electrodes tend to fail after a while because they get gummed up by the body's immune reactions, and when they do fail, they necessitate a second surgery to have them removed. As a result, many researchers have been concentrating their efforts on improving the effectiveness of externally worn, noninvasive devices such as skullcaps and headsets—with results that have proved nothing short of astounding.

Go to YouTube again and type in "Berlin brain computer interface."[11] The year is 2006. A young man sits in front of a computer, wearing what looks like a bathing cap with dozens of small electrodes protruding from it. All around him, we see people passing by, milling about: he is a walk-on volunteer at a booth at a major German conference on brain-machine interfaces. A few minutes earlier, this young man had sat down, and a booth attendant had strapped on the cap and explained what to do: think of your right hand for "up," your left hand for "down." It took a few minutes to calibrate the computer, so that the electroencephalograph (EEG) signals it was getting from the skullcap could be correlated with intended movements on the screen.[12]

That's all it takes. The young man with the skullcap sits motionless, staring intently at the video screen, and begins playing Pong, controlling the paddles on the screen with nothing but his thoughts. At first, he is pretty bad at it, but over the following minutes, his performance steadily improves, until he is moving the paddles almost as well as he would with his hand on a mouse or keyboard. When this young man grows tired of the game, he can simply take off the cap, clean the conductive gel from his hair, and get on with the rest of his day. No surgery, no implant, no fuss.

Scroll forward to the present day, and you can buy a similar brain-computer-interface device to play with at home. This time, though, it's not a geeky-looking skullcap; nor is there any need for conductive gel. You just slip on a stylish little headset, and you're good to go. The device is called MindWave, and you can buy it for $129.[13] The headset uses EEG sensors to assess your fluctuating levels of concentration and harnesses this information to give you basic controls for interacting with informatic devices ranging from computers to cell phones.

The sheer variety of these noninvasive systems has grown explosively over the past twenty years.[14] Some skullcaps and headsets measure tiny variations in cerebral blood flow; others tune in on magnetic fields; others monitor brain waves and electrical activity. Some primarily monitor activity near the brain's convoluted exterior layers, while others attempt to listen deep below the surface. At the same time, researchers are running a parallel race to refine the mathematical sorting algorithms they employ to tease useful information from amid the torrent of data the sensors pick up.[15]

Restoring memory and cognition

The goal, admits the biomedical engineer Theodore Berger, with exquisite understatement, "may initially appear somewhat daunting."[16] Indeed it may. Over the past twenty years, he has been working with an international team of researchers in neuroscience, cognitive psychology, molecular biology, mathematics, biomedical engineering, computer science, electrical engineering, and materials science to build what he calls "replacement parts for the brain." (This is not just an off-the-cuff phrase: it is the title of a book coedited by Berger in 2005, with contributions from dozens of luminaries in all these fields.)[17] Berger and his coworkers are well on the way toward creating a neural prosthesis for the human hippocampus, the small seahorse-shaped organ deep within the brain that plays a key role in converting short-term experiences into long-term memories.

When things go wrong with the hippocampus, as they sometimes do with Alzheimer's disease and certain kinds of brain trauma, things also tend to go very wrong with patients' basic sense of who they are. Such patients become unable to form declarative memories, the kinds of memories that tell us what the state of the world around us is over time. If scientists could find a way to repair a malfunctioning hippocampus with a neural prosthesis, the impact on thousands of people's lives would be incalculable.

In 2006 Berger and a group of colleagues performed one of the landmark experiments in the nascent field of neural prosthetics.[18] They developed a microchip that could accurately reestablish electrical processing across gaps or lesions in the hippocampal tissue of lab rats. By 2011 Berger and his colleagues had successfully tested the device in live rats, demonstrating that the hippocampal prosthesis could be used to selectively activate and deactivate the rats' ability to form long-term memories. "Flip the switch on," Berger noted, "and the rats remember. Flip it off, and the rats forget."[19] The device could be used, moreover, not only to replace the tissues of a damaged hippocampus, but also to boost the performance of a normally functioning hippocampus. "For the first time," Berger reported, "a neural prosthesis capable of real-time diagnosis and manipulation of the encoding process can restore and even enhance cognitive, mnemonic processes."[20] Berger believes that, somewhere around the year 2025, the first human trials of a hippocampal prosthesis could well be taking place.[21]

Monitoring the contents of the mind

In 1999 a team of neuroscientists at the University of California, Berkeley, reconstructed the images that a cat was seeing, just by reading the electrical activity of neurons in its brain. They placed electrodes into the visual circuitry of the cat's brain, retrieving the activity of neurons while the cat was being shown photos of natural scenes such as a tree branch or a human face. When the information stream from the electrodes was decoded, the results were remarkable: a blurry, but still recognizable reproduction of the original photo that had initially been presented to the cat.[22]

In 2008 the same feat was replicated with humans, this time using non-invasive fMRI technology.[23] A team of researchers at Japan's ATR Computational Neuroscience Laboratories in Kyoto placed human test subjects inside an MRI machine and then showed them repeated sets of images of the letters of the word "neuron." The MRI machine homed in on the subject's visual cortex, measuring the shifts in electrical activity there, and then transmitted the data to a computer that in turn translated it into a ten-by-ten pixel array on a video monitor.

The result is somewhat eerie to contemplate. We see eight rows of black-and-white images, each bearing a blurry, pixelated array of dots in which we can faintly recognize the outline of a letter: n-e-u-r-o-n. At the bottom, the computer also generates a final array that computes the average of all the other readings: here the word "neuron" pops out much more crisply. I know as I look at it that I would be able to read the word without much difficulty, even if I had not been told what it was in advance. I am seeing an image that has been extracted by a machine directly from a person's conscious mind.

Altering states of mind and emotion

One of the most famous experiments in postwar biology took place in the mid-1950s, when the neuroscientist James Olds gradually mapped out the brain's "pleasure center" in rats.[24] Olds implanted permanent electrodes into the rats' brains and rigged a lever mechanism so that it would trigger a mild electric pulse into the electrodes whenever the rat pressed it. The animal, in essence, had been given the capability of electrically stimulating its own brain on demand. When Olds targeted the electrodes to certain areas in the midbrain, the rats began frenziedly pressing the lever again and again as if their very lives depended on it. What was more, he discovered, the rats preferred the self-stimulation of these pleasure centers to all other behaviors or stimuli. They ignored nearby food and water, even after intervals of twenty-four hours had gone by since they had last been fed. Instead, the animals just stood before the lever and continued pressing on it until eventually they keeled over from sheer exhaustion.

Were the rats really feeling pleasure? Subsequent experiments with humans suggested that they were indeed. A pair of Norwegian researchers reported in the late 1950s that when they placed electrodes into certain areas of their patients' brains associated with pleasurable sensations and then allowed the patients to self-stimulate using a push-button device, "some subjects actually stimulated themselves into convulsions."[25] A few years later, in the 1960s, the Tulane University neuroscientist Robert Heath began using a similar method on some of his own patients, who were suffering from severe and intractable depression.[26] One of his test subjects, whom he called B-19, had electrodes implanted deep into a variety of regions of his brain. "The patient," Heath reported, "was allowed free access to the buttons of the self-stimulator which activated the left anterior and right midseptal leads. . . . During these sessions, B-19 stimulated himself to a point that, both behaviorally and introspectively, he was experiencing an almost overwhelming euphoria and elation and had to be disconnected, despite his vigorous protests."[27]

In the early 1970s Heath went on to develop a portable, pacemaker-like version of his brain-stimulation apparatus and implanted it subcutaneously in a number of patients. Here is an account of one patient's experience.

The first patient to receive the [portable] implant was a young man who flew into uncontrollable, violent rages and had to be tied to his bed. A small hole was drilled into his skull, and the tiny electrode was slipped into his brain and fired to activate the pleasure center—located in the septal area and in part of the amygdala—and to inhibit the place where rage is centered—the other part of the amygdala, the hippocampus, the thalamus, and the tegmentum. The unit delivered five minutes of gently pulsed current every ten minutes and worked quite well. The man was untied, let out of his bed, and eventually allowed to go home. Things went peaceably for a while. Then one day the man went on a rampage in which he wounded a neighbor, attempted to murder his parents, and barely escaped being shot by the police. He was carted back to the institution and x-rayed. The wires from the battery to the [brain-stimulation] pacemaker had frayed and broken. Heath reattached the wires, and the young man calmed down and went home again.[28]

One would think that these kinds of remarkable achievements with brain implants would have paved the way for a major research effort, coupled with a vigorous campaign by manufacturers of medical electronics to develop implantable devices for the wider market. But this was not to be. Following the 1970s, the field entered a period of relative doldrums from which it did not begin to emerge again until the early 2000s. The reason is probably twofold. On one hand, the level of understanding that neuroscientists had of the brain's functioning remained so rudimentary in the 1970s and 1980s that most scientists and doctors simply considered these kinds of brain implants excessively risky. Heath, in this sense, was something of an outlier, willing to take chances that other doctors avoided.

At the same time, the political and moral climate surrounding medical research was rapidly changing. Bioethical considerations about the rights of animals, and about the welfare, dignity, and informed consent of human subjects, gradually came to play a more prominent role in the shaping of experiments and medical interventions. Government regulation grew more stringent, and universities established institutional review boards to ensure that their researchers were adhering to an increasingly comprehensive set of ethical guidelines.[29] Researchers in brain science therefore shied away from the kinds of experiments that Heath had done, focusing their attention more on animal studies, circumspect work with human subjects, and the fundamental neuroscience, molecular biology, genetics, and cognitive psychology that underpinned the brain's functioning.[30]

Direct modulation of brain activity

Four types of therapeutic technologies, meanwhile, are allowing researchers to penetrate the human brain and modulate its electrochemical activity in increasingly precise ways. A technique known as deep brain stimulation (DBS) was first developed by French physicians in the 1980s as a treatment for severe Parkinson's disease.[31] As the name implies, it involves the implantation of permanent electrodes into a variety of regions well below the surface of the brain. The electrodes are attached to a pacemaker-style implant that delivers electric current on a programmable schedule. DBS has proved surprisingly effective in treating a variety of conditions, including not just Parkinson's disease but also Tourette syndrome, severe depression, obsessive-compulsive disorder, and intractable pain.[32]

Another technology currently under development is transcranial magnetic stimulation (TMS), which involves targeting highly focused magnetic fields at specific regions deep within the brain through an electromagnetic coil held against the forehead. Such fields are known to produce effects similar to direct stimulation by implanted electrodes, but they have the considerable advantage of being noninvasive in nature.[33] This technology remains experimental today, but researchers believe it may come to play an increasingly important role in brain interventions over coming decades.[34]

A third promising field, in this regard, is optogenetics—a cutting-edge method that uses light signals to modulate the firing of individual neurons in a living brain. Ordinarily, neurons are unaffected by light, but if they can be induced to become photosensitive, then their firing rates can be controlled on a millisecond-by-millisecond basis by shining bursts of light onto them through fiber-optic micro-wires.[35] One of the field's leading practitioners, MIT's Edward Boyden, believes that if scientists find a way, through targeted gene therapy, to turn certain portions of a person's brain tissue photosensitive—without generating harmful side effects—then optogenetic devices could someday deliver extraordinary tools either for curing neurological diseases or for directly modifying various forms of mental function.[36]

Last but not least, of course, there is the good old technique of simply hooking up electrodes to your temples and zapping you. Scientists (and quacks) have been trying this out on patients ever since the mid-nineteenth century, with few reliable results to report—until recently.[37] The procedure, known today as transcranial direct current stimulation (tDCS), is being studied primarily as a method for dramatically improving human cognitive performance.[38] The test subject sits in front of a computer console, tasked with acquiring a complex new set of skills in spatial recognition. An electrode array delivers two milliamperes of direct current over a thirty-minute period to the cortical areas beneath her temples and forehead. "It feels like a mild tingling or a slight burning," reports one of the test volunteers.[39] To the astonishment of researchers, learning rates under tDCS increase significantly, sometimes cutting the time required for full acquisition of a new skill by up to 50 percent. "By using electricity to energize neural circuits in the cerebral cortex," observes a reporter for *Scientific American*, "researchers are hopeful that they have found a harmless and drug-free way to double the speed of learning."[40]

WHERE WE MAY BE AT MIDCENTURY

Let us start with four general principles that will probably characterize the trajectory of bioelectronics over coming decades.

1. *Growing powers, increasing seamlessness.* Our understanding of the brain, sensorium, and nervous system will continue to deepen. Medical technologies will become more sophisticated—ranging from better surgical procedures to more seamlessly integrated implantable devices of all kinds. Our ability to link living tissue and artificial devices will grow impressively. Computers will become smaller and more powerful, and their software will be more interactive and better adapted to eliciting and responding to human creativity. Robotic devices will become more

flexible, independent, useful, and ubiquitous. Prostheses of all kinds will incorporate tactile and sensory feedback capabilities. The overall trend, in short, will be for our increasingly powerful machines to mesh ever more intimately with their human operators.

2. *Noninvasive devices will predominate.* In the long run, the market for noninvasive bioelectronic devices will prove much larger than the one for invasive implants, particularly when the noninvasives begin to enter common use in computer games and virtual reality programs (already a billion-dollar business today). Once that happens—probably by the year 2025, with early applications arriving even sooner—it is likely that vast quantities of corporate funding will flow toward skullcap technologies and other noninvasives.[41] This influx of money and attention may well, in turn, lead to the invention of still more potent and effective interfaces whose nature we cannot even envision at present.

3. *Treatment unavoidably opens the door for enhancement.* All the bioelectronic devices we have surveyed open up tangible possibilities for human enhancement. Why? Because the technologies for repairing a malfunctioning human body are inseparable from the technologies that allow us to push human capabilities to ever higher levels. To the extent that bioelectronic devices can restore lost function in the disabled or ill, they can also be adapted at some point down the line to boost the functioning of healthy people, or to give those people entirely new powers that no human has ever possessed. Over time, bioelectronics designed for therapeutic applications will probably come to occupy a relatively small market niche compared with the dominant strain of entertainment- or enhancement-oriented products.[42]

4. *Beware the hype.* Some journalists are prone to making wild leaps, taking an important but limited technological advance and inflating it into a howling sci-fi attention grabber. For example, when Honda researchers succeeded in 2009 in getting their ASIMO robot to respond to four preprogrammed commands transmitted from a technician via bioelectronic skullcap, the corresponding headline in *Businessweek* announced: "From Honda, a Mind-Reading Robot."[43]

This kind of silly hype is easy to spot and dismiss with a chuckle. Another variety, however, is more insidious. It is particularly prevalent when we speak of using bioelectronics for augmenting the higher cognitive or emotional functions. Thus, for example, when we refer to Berger's hippocampal prosthesis as a "memory-boosting chip," this terminology is in one sense perfectly legitimate. The chip replicates neural activity in a part of the brain that is central to processing short-term memories into long-term memories. To most people, however, the notion of "boosting memory" carries connotations of something quite different. It implies being able to read an encyclopedia once over and, three days later, to summon up with perfect accuracy the entry on page 355. Or being able to reexperience at will the exact flavor of the salmon mousse I ate on that picnic with my girlfriend back in Marin County, California, on June 11, 1972. This, of course, is misleading. Not only does the technology under development come nowhere near this level of potency, but it is not yet clear whether *any* technology will ever be able to offer such a capability. In discussing the future of bioelectronics, therefore, we should retain a certain skepticism and humility. In the words of the neuroscientist Philip Kennedy: "We don't even know what a thought is yet."[44]

What, then, might some of the bioelectronics of the year 2040 or 2050 actually look like?

Sensing

Our five senses will probably be linkable to machines in startling new ways. Alongside visual-system enhancements, for example, we can imagine using a cochlear sound augmenter to hear what dogs and cats hear, or what bees and moths hear, tapping directly into their sonic universe for a visit to see what's up. A tactile enhancement technology would offer extensions of our body through touch-sensitive haptic devices, allowing us to manipulate objects in outer space or on another continent, while "feeling" them as if they were at arm's length.[45]

Still more exciting would be the incorporation of entirely new *kinds* of sensations that we have a harder time imagining today. Consider the following experiment, described in a recent *Wired* article:

> For six weird weeks in the fall of 2004, Udo Wächter had an unerring sense of direction. Every morning after he got out of the shower, Wächter, a [technician] at the University of Osnabrück in Germany, put on a wide beige belt lined with 13 vibrating pads—the same weight-and-gear modules that make a cell phone judder. On the outside of the belt were a power supply and a sensor that detected Earth's magnetic field. Whichever buzzer was pointing north would go off. Constantly . . .

The effects of the "feelSpace belt"—as its inventor, Osnabrück cognitive scientist Peter König, dubbed the device—became even more profound over time. König says while he wore it he was "intuitively aware of the direction of my home or my office. I'd be waiting in line in the cafeteria and spontaneously think: I live over there." On a visit to Hamburg, about 100 miles away, he noticed that he was conscious of the direction of his hometown. Wächter felt the vibration in his dreams, moving around his waist, just like when he was awake.[46]

One animal species capable of sensing the earth's magnetic field is the loggerhead turtle. Even small hatchlings possess an uncanny ability to feel the orientation and strength of magnetic fields with great accuracy: they use this ability to guide them through their eight-thousand-mile annual migrations across the seas.[47] The belt worn by Wächter and König—a fairly simple and noninvasive technology—resulted in a partial remapping of the sensory space they felt themselves inhabiting, moment to moment. Even their dreams came to be tuned to North. Tasting, touching, seeing, hearing, smelling—*and orienting*. They had successfully added a new "magnetic" dimension to their sensorium and were thereby able to share the kind of sensory space occupied by sea turtles. It is possible that the bioelectronics of the mid-twenty-first century will allow us to broaden our sensory sphere in analogous but even more potent ways.

Controlling machines

We started the computer age with punch cards, magnetic tape, and typewriter keyboards. Then, in 1984 (!), Apple introduced the Macintosh, with its mouse and graphic user interface, dramatically transforming the way humans and computers could interact. Today, researchers are developing interfaces that allow us not only to converse directly with our machines using spoken language (How ya doin', Siri?), but also to interact with robotic devices that have highly expressive *faces* capable of communicating complex states such as curiosity, satisfaction, or surprise.[48]

What other forms of interaction would go beyond the rich world of spoken language? One possibility would be a computer-generated, interactive, virtual environment projected for me—a sphere of objects and symbols that I could manipulate at will using the two-way communication channel of a skullcap interface. In lieu of encountering the machine by sitting in front of the monitor as I do today, such an interface would situate me inside a three-dimensional "place" that I would temporarily inhabit. This interactive space could perhaps offer me an ability to imagine a visual image or a situation, and have the machine instantly pick it up, project it, and start engaging me about it. In this sense, the machine would be connecting with my mind and sensorium on a much broader spectrum than today, allowing me far greater degrees of freedom in my interactions with it.[49]

Augmenting cognition

Could a neural prosthesis render us smarter or allow us to communicate directly with each other, brain to brain? In the film *The Matrix*, we encounter a tantalizing vision of such a technology. The character named Trinity finds herself urgently in need of the ability to fly a helicopter; she calls her buddy on a cell phone, and with a few clicks on his computer, he transmits directly into her brain the know-how of an experienced pilot; her eyes flutter for a couple seconds, then she looks at her friend Neo and says, "Let's go."[50] The information transmitted here involves something that cognitive psychologists call procedural knowledge—the "how" of doing things, like riding a bike or making a really mean cheese soufflé. What Trinity downloads, in other words, must not merely be an owner's manual for an M-109 helicopter. She is in a hurry and needs to fly the damn thing now. She must acquire, in three or four seconds, the accumulated experiential knowledge that someone would have incrementally gained over many long sessions in flight school—assimilating the reflexive body motions, the technical information, the theory and basic understanding, and, above all, the training required to operate a complex machine.

This kind of knowledge certainly does exist in the brains of trained pilots, and it is therefore theoretically possible that, someday, we may be able to single out and record the neural patterns that underpin this intricate set of operational abilities. Translating and adapting those neural patterns within the context of a different person's brain (and body) would no doubt constitute an equally daunting challenge. I'll discuss these possibilities in greater detail later.

THE *TERMINATOR* FALLACY (PARANOID COUSIN OF THE *JETSONS* FALLACY)

The story line of the popular *Terminator* films follows one of the most venerable tropes in the history of science fiction: intelligent machines taking over, super-powerful robots running amok.[51] It makes for an alluring story, particularly in an age of rapid mechanization, and variants of this plot have animated movies and novels starting with Fritz Lang's Expressionist masterpiece of 1927, *Metropolis*, on up to modern films like *2001*; *I, Robot*; and *Transformers*.[52] But such an image of the middle-term future is probably misleading. The machines will not be our enemies, standing against us in a clear battle between us and them. They will not take over. Instead, they will become part of us. We will invite them progressively not just into our professional and personal lives (which we have already done to a considerable extent) but into our very bodies and minds. The trajectory, in short, is one of steady convergence of humans and machines, as we (literally) "in-corporate" them into the fabric of our identities as individuals. The machines will never be able to "take over" in this sense, because we will already have partially merged with them. Voluntarily. Eagerly.

It sounds creepy at first to put it this way, but past experience with how humans have adapted to new technologies suggests that many people, over coming decades, may not actually perceive it as creepy at all. A key concept here is what cognitive psychologists call "transparency"—the uncanny ability that humans have to remap the boundaries of their body image, integrating various external devices into the natural functioning of their movements and thoughts. As I drive around town in my car, for example, I do not have to say to myself, "OK, that's a red light, so I need to lift right foot, place it on brake pedal, apply gradual pressure, wait for car to slow down, now let up on the pressure as the car has fully ceased to move." Instead I just know that I am driving, and I think, "Red light, stop." Over time, the machine interface has come to be integrated into my own body image, so that operating the device feels as though I were simply willing a movement of my own limbs.

My experience of transparency with an automobile is constrained by the fact that I can easily distinguish my selfhood from the machine. When I get to my destination, I exit the vehicle and go about my business. Self here, car there. But what will happen when—as with Jens Naumann and his visual prosthesis—the machine becomes an integral part of my body's functioning? How much deeper will the effect of transparency run when the device stays with me all the time, and the capabilities it offers become a basic element in my sense of who I am? (Think of how you feel when you lose your cell phone or the power goes off in your house.) Once our lives are lived in a physical and virtual space that is permeated by the machines' co-activity, it is reasonable to expect that our level of identification with them will go sharply up. We will live increasingly as a continuum of overlapping machine/human sensations and undertakings.

This will not happen overnight. The transition will take place slowly, in uneven increments of innovation that will accumulate over many years. It is possible (though by no means certain) that our society will adapt with relative smoothness to these changes, since it will be we ourselves who would presumably be choosing them at every step along the way. At the same time, however, we should be under no illusions. By the middle decades of this century, a revolutionary transformation will have taken place. Novel forms of sensation, powerful prostheses, sophisticated robots and drones controlled by thought, interactive knowledge-spaces following our movements, radically new modes of interpersonal communication—these are only the most likely and foreseeable extensions of our bioelectronic selves.

FOUR

Genetics and Epigenetics

The horizons of the new eugenics are in principle boundless—for we should
have the potential to create new genes and new qualities yet undreamed. . . .
For the first time in all time, a living creature understands its origin and can
undertake to design its future.

—Robert Sinsheimer, molecular biologist, chancellor,
University of California, Santa Cruz (1969)[1]

In the history of genetic interventions, there have been four major shifts that
I'd like to explore, beginning, of course, with Mendel.

FROM MENDEL TO DOLLY THE SHEEP

Scientists use the word "genotype" to refer to an organism's ensemble of
DNA code, the creature's full set of hereditary information. It stands in contrast to the "phenotype," the totality of actual traits observable in the organism as a whole.[2] A gene is a unit of heredity, corresponding to a specific
segment of DNA code.[3] It can best be thought of as a finite set of instructions that tells the body how to go about the basic functions it requires,
either for building the tissues of a developing organism or for maintaining
the life processes of a fully developed organism.

In Gregor Mendel's day, the practice of altering species of plants and
animals was conducted on a hit-or-miss basis, by observing certain desirable
traits in parent organisms and then interbreeding those creatures and waiting to see if the desired traits would be expressed in a more pronounced way
in the offspring. Today, by contrast, scientists can engineer specific traits
into plants and animals by directly inserting new genes into their DNA.
They can snip out the green-fluorescence gene from the DNA of a jellyfish,
for example, and splice it into the DNA of a rabbit embryo. When the rabbit
is born, voilà, a bunny that glows green under a specific wavelength of blue
light.[4] More usefully, scientists can extract a gene from the bacterium *Bacillus
Thuringiensis* that codes for the secretion of a chemical that is toxic to certain

classes of insects. When this gene is inserted into the DNA of crops such as corn, potatoes, and cotton, it causes all successive generations of the plant to manufacture the bacterial toxin within its own tissues, thereby rendering the crops resistant to major insect pests. Sixty-three percent of the corn planted in the United States in 2010 contained this genetic modification.[5]

The significance of this shift in capabilities can hardly be overstated. It is like the difference between groping one's way across a pitch-dark room, as compared with striding purposefully through a brightly lit space. Modern genetic science takes the slow, trial-and-error processes of evolution and natural selection, and transforms them into a new kind of directed evolution more akin to manufacturing or art. Human bodies and minds, of course, fall squarely within the purview of these nascent capabilities for partial alteration or wholesale redesign.

EPIGENETICS: OPENING UP NEW PATHWAYS FOR HUMAN ENHANCEMENT

Over the past half century, the rise of molecular biology has transformed our understanding of how genes operate. In this emerging picture, genes do a lot more than just determine the inheritance of traits: they remain active agents in the ordering of an organism's basic life functions, from cradle to grave.[6] They help regulate the ongoing biochemical processes inside every cell; they interact with hormones to turn up or turn down the secretions of chemicals that govern major organ systems; they mediate the cascade of chemical signals that allows the immune system to respond to toxins and other foreign bodies; they respond to chemical stimuli in the brain, modulating the performance of this most important of all organs. Genes, in other words, are active participants not just in making us who we are, but in keeping us alive and healthy moment after moment.[7]

The more we learn about the functioning of genes, the more salient becomes the role played by the mediating factors that regulate their activation, deactivation, and transcription. Many of our twenty thousand genes, it turns out, are constantly being switched on and off in complex combinations, as well as in still more complex chronological *sequences* of combinations. The activity is profoundly bidirectional, in the sense that genes both regulate, and are regulated by, the chemical processes going on in the cell and broader organism that surrounds them.

The new scientific field that studies these patterns of genetic activation and transcription is known as epigenetics.[8] Though definitions vary, an epigenetic process can best be described as any molecular mechanism that changes the expression of genetic information without altering the underlying DNA sequence itself. The DNA code stays the same, but certain portions of it are selectively silenced, while others are spurred to action, resulting in dramatically different phenotypic outcomes. The biologist Nessa Carey explains it this way:

> We talk about DNA as if it's a template, like a mold for a car part in a factory. In the factory, molten metal or plastic gets poured into the mold thousands of times, and, unless something goes wrong in the process, out pop thousands of identical car parts.
>
> But DNA isn't really like that. It's more like a script. Think of *Romeo and Juliet*, for example. In 1936 George Cukor directed Leslie Howard and Norma Shearer in a film version. Sixty years later Baz Luhrmann directed Leonardo DiCaprio and Claire Danes in another movie version of this play. Both productions used Shakespeare's script, yet the two movies are entirely different. Identical starting points, different outcomes.
>
> That's what happens when cells read the genetic code that's in DNA. The same script can result in different productions.[9]

In recent years, scientists have discovered a variety of epigenetic mechanisms that allow the DNA script to be read differently by the body's cells under distinct circumstances; the two most common of these are known as DNA methylation and histone acetylation.[10] These two molecular mechanisms act like volume knobs on particular segments of DNA: one mechanism (methylation) turns down the potency of expression for a given section of code, all the way down to a whisper; the other (acetylation) cranks it up to a shout. (Although we often speak of genes being turned on or off, like switches, they operate in most cases more like analog knobs on a dial, varying the potency of their expression by increments and degrees.) This molecular control process, biologists believe, is what allows the cells in your body to differentiate into skin cells, hair cells, nerve cells, and so on, even though every one of them has exactly the same underlying DNA code in its nucleus. The core script is identical in every case, but the way the epigenetic mechanisms cause the script to be read determines which qualities and functions each cell will come to possess.

What is more, these precise epigenetic patterns for regulating DNA expression are passed on from each cell to both of its daughter cells at the moment of cell division, thereby ensuring that a skin cell continues to replace itself with more skin cells rather than nerve or muscle cells. Scientists have therefore added an important new term to our vocabulary: just as the sum total of your DNA information is known as your genome, the sum total of these molecular "script-reading" instructions is known as your epigenome. Together, perpetually locked in an intricate pas de deux, the symphonic interactions of your genome and your epigenome run the biochemistry of the organism that is your body.[11]

This emerging model of genetic/epigenetic causation has major implications for human enhancement. If scientists figure out how to control and modify these epigenetic mechanisms, this will give them a powerful new tool for genetically reshaping all manner of biological organisms, without having to make the slightest alteration to the creature's underlying DNA. Epigenetic modifications would be considerably easier to make and would result in flexible phenotypic changes that could be further altered, upgraded, or even reversed at a later time. I'll come back to this in greater detail later.

WHICH TRAITS ARE SUSCEPTIBLE TO GENETIC MODIFICATION?

Behavioral geneticists tend to distinguish three main factors that weigh heavily in shaping our physical and mental traits: the genetic component, the shared family environment, and the nonshared environment outside the family (such as school, friends, our unique set of experiences.) The balance among these three elements—their relative degree of shaping power—differs significantly from one trait to another.[12] The fact that I am a male, for example, was straightforwardly determined by my genes: even if I had been reared under dramatically different conditions, I would still remain biologically male. On the other hand, the fact that I speak English as my primary language, rather than French or Chinese, has nothing to do with my genes: it was entirely determined by the environment in which I was raised.

In many cases, it is misguided to speak of a "gene for" a specific trait, the way the mass media tend all too often to do. A gene for intelligence, for homosexuality, for shyness, for criminality—to a geneticist, these are all nonsensical ideas.[13] Such complex human traits arise from the interaction of many individual genes and suites of genes over long periods of time, from the interaction of those genes with a variety of epigenetic regulators and transcription chemicals, from the interaction of both genes and regulators with all manner of cellular environmental factors as the organism develops, and from the interaction of the organism as a whole with larger-scale environmental factors such as climate, socioeconomic milieu, or cultural context. To speak of a "gene for" one of these traits, in the face of such multilayered complexity, is a bit like speaking of the "average sound" of a Beethoven sonata. It simplifies to the point of absurdity.

The greatest source of misunderstanding here lies in confusing the role played by genetic causation in the case of a single individual, as compared with its role in determining *rates* of a trait's manifestation across broad segments of a population. As a statistical concept, applied to large numbers of humans, the notion of a quantifiable genetic contribution to a given trait does make sense. Behavioral geneticists seek to measure the *heritability* of different traits in a given population, by which they mean "how much of the total phenotypic variation in a trait can be explained by genetic variation alone."[14]

Some personal characteristics, such as sense of humor or food preferences, show low susceptibility to genetic influence: in twin and sibling studies, researchers have discovered that adopted siblings (who share no familial DNA) nonetheless tend to show similar food preferences and senses of humor, whereas monozygotic (or identical) twins reared apart develop divergent ones.[15] These traits, therefore, are more decisively influenced by the home environment than they are by genetic factors. The same goes for political attitudes and religious affiliation: these correlate strongly with the beliefs that predominate in the home environment, suggesting that the cultural influence of the family milieu weighs more heavily than the individual's DNA in determining them.[16]

The situation is quite different, however, with cognitive abilities and the features of personality. Behavioral geneticists have determined that about 40 percent of the variation in personality traits such as affability, character, or shyness is attributable to direct genetic factors; about 10 percent is linked to the shared family environment; and about 25 percent comes from environmental factors outside the family (the remaining 25 percent is attributable to measurement error).[17] Cognitive traits such as IQ are also heavily influenced by genetic factors: here the genetic component is about 50 percent, the shared family environment counts for about 25 percent, and the nonfamilial environment influences the remaining 25 percent.[18]

What are we to make of these kinds of findings? Do they suggest a form of genetic fatalism? Far from it. The fact that genetic factors account for about 40 percent of the variance in people's body weight, for example, says nothing about how fat I myself will become.[19] It merely indicates that people's *propensity* to become fat or thin runs to a considerable extent in their bloodlines: some people (whom I deeply resent) are able to eat like horses and yet remain effortlessly thin nonetheless. For myself, I can put to use this finding of behavioral genetics in the following way. Having observed that my parents and grandparents are all paunchy, and plugging in the 40 percent heritability statistic as a probabilistic factor, I reach a grim conclusion: I have a significant chance of being among those portions of the population whose genes predispose them toward excessive gravitas. I cannot know this for sure, of course: my application of these statistics to my own case remains nothing but an educated guess. Still, it induces me to launch a program of regular exercise and a low-fat diet. Far from reducing me to a state of fatalistic apathy, my understanding of the statistics of genetic propensities empowers me. The ultimate decision about what to make of myself remains solidly my own.

This conclusion applies to the more ethereal traits as well. In all the findings of behavioral genetics, it is clear that both nature and nurture play pivotal roles in shaping who we become. One portion of the basic parameters is set by our DNA inheritance, but the eventual phenotypic outcome can be powerfully wrenched toward one direction or another by our experiences (and choices) after birth. You can take the child of a family of geniuses, whose DNA might predispose her to rank fairly high on the IQ scale, and drastically dumb her down by rearing her in an impoverished intellectual environment.[20] You can take a child from a family of seven-foot-tall basketball players and stunt his growth dramatically through poor nutrition. On the other hand, there are limits to how far environmental factors can go. If a child is born to parents who are both five feet tall, it is likely that no amount of milk will turn him into a seven-footer. If you want him to grow that tall, you will need more potent biochemical tools, such as heavy doses of human growth hormone administered throughout his youth—and even then the outcome will probably not be a LeBron James.

EUGENICS

The project of "making better people" gained widespread acceptance in many nations during the first half of the twentieth century; the United States in particular became a world leader in the implementation of state policies for "improving the stock."[21] Eugenic practices came in two forms. Negative eugenics consisted in discouraging, through a variety of means, the reproduction of "undesirable" types of persons. (Which traits counted as "desirable" tended, not surprisingly, to reflect the particular background of the person doing the eugenic judging: North European, middle-class, white, male qualities tended to predominate.) Positive eugenics, on the other hand, entailed increasing the reproductive rates of "desirable" categories of people.

Eugenic ideas reached their culminating expression in the practices adopted by the Nazi regime during the 1930s and 1940s—selective breeding, forced euthanasia, and the death camps.[22] After 1945, as the world recoiled in revulsion at the Nazi crimes, eugenic thinking all but disappeared from public discourse. Its association with the Hitler regime had exposed most vividly the harsh and inhumane premises on which it rested. But the underlying eugenic impulse—the lure of building better humans—turned out to be far from dead. It lay dormant for several decades and then gradually began to make a comeback in a dramatically different form.[23]

The new version of eugenics starts off with a vehement denunciation of the old. Its proponents argue that no government should ever have the power to dictate which human traits are desirable and which are not. But it is not the *project itself* of improving human bodies and minds that is the problem, according to this view: rather, it is a question of who decides, a question of freedom and choice.[24] If I am offered technologies that let me alter or redesign my own physical and mental constitution, then I should be free to undertake or avoid such alterations as I wish, as long as these alterations do not directly harm anyone else. This is the new incarnation of eugenic thought, which one scholar has aptly dubbed "liberal eugenics."[25] To the extent that the human species will be deliberately reshaped over time, the liberal eugenicists believe, this will take place through the "invisible hand" of myriad decisions made by private citizens acting on behalf of their own values, worldviews, and perceived interests.

TODAY'S STATE OF THE ART IN GENETIC INTERVENTIONS
Genetic engineering of plants and animals

Only a small percentage of scientists' genetic interventions have thus far taken place with human patients: the vast majority of recombinant experiments and procedures have focused on the plant and animal world.[26] Some of these interventions have involved *somatic* alterations, which merely modify the genetic information in certain tissues of a creature, but which are not passed on to subsequent generations when that creature reproduces. Other interventions have brought about a far more potent *germline* alteration: by modifying the DNA in a creature's sex cells (ovum and sperm in mammals, for example), this kind of intervention causes changes that continue to be passed on through all subsequent generations of a creature's descendants.

Over the past twenty years, scientists have not only engineered crop plants for resistance to a wide variety of pests, but have also used recombinant techniques to turn other plant and animal species into veritable factories of useful chemicals. One species of tobacco plant, for example, has been modified to churn out human glucocerebrosidase in its leaves—an enzyme crucial to treating persons with Gaucher disease.[27] Transgenic goats now secrete the human blood-thinning protein antithrombin III in their milk, a modification that allows production of the protein for roughly one-hundredth the cost of traditional cell-culturing procedures.[28] All these forms of genetic modification have generated fierce controversy.[29] Critics have raised a variety of concerns, ranging from fears of environmental damage to worries about unintended health effects. Defenders of genetic engineering, for their part, have emphasized the significant potential benefits of the technology, from curing diseases to alleviating poverty.[30]

Cloning: the science and the silliness

Two kinds of cloning exist today.[31] The first, molecular cloning (also known as recombinant DNA technology), has been in widespread use since the 1970s: it is the technique that allows scientists to isolate and copy specific segments of DNA for sequencing, study, and manipulation. The second kind of cloning involves a delicate and technically arduous process known as somatic cell nuclear transfer. Scientists remove and discard the nucleus from an egg cell, then introduce into the egg cell a new nucleus taken from a body (or somatic) cell belonging to a different creature. The resultant hybrid cell then begins dividing and multiplying, following the coded genetic instructions contained in the somatic cell nucleus that has been implanted into it. After a few days it forms a multicelled blastocyst, or early embryo.

At this point, two possibilities emerge: if the blastocyst is implanted into the uterus of a surrogate mother, it may develop into an embryo and eventually be born as a nearly identical genetic copy of the donor creature whose somatic cell provided the new nucleus.[32] This is known as *reproductive cloning*: it is the process that created Dolly the sheep in 1996, the first mammal successfully cloned from an adult somatic cell. On the other hand, scientists can also opt not to implant the blastocyst into a surrogate mother, but instead to culture the cells in the laboratory for research purposes. This is known as *therapeutic cloning*, because scientists hope to use the technique not to create a full-grown organism but rather to create human stem cells, which hold great promise in a wide range of medical applications.[33]

Few subjects in biotechnology have generated as much emotional Sturm und Drang as the prospect of human reproductive cloning.[34] In the *Star Wars* franchise, one sees vast legions of indistinguishable clone troopers on the march; on the cover of magazines, one sees images of identical babies proliferating across the page. Much of the excitement, alas, is based on simple ignorance. Large numbers of people in our society are unwitting genetic determinists: they continue to think that a clone of me would not only look like me, but would eventually grow up to have the same tastes, disposition, talents, and shortcomings—even the same thoughts.[35] No matter how hard scientists try to disabuse people of this mistaken belief, it continues to linger, deeply embedded in the culture.

In reality, as the owners of cloned pets have discovered, the interplay of genes and environment renders it impossible for a clone to be phenotypically identical to its parent organism.[36] At most, a human clone would be somewhat like a monozygotic twin reared apart from its sibling, bearing many striking similarities to its clone parent as well as many equally striking differences. The clone armies in *Star Wars*, if depicted realistically, should show individual troopers going every which way, each marching to his own unique inner drummer. (Parenthetical note: a glance at the history of Germany and Japan in the 1930s shows that you don't need cloned humans in order to produce phalanxes of fanatical warriors marching in lockstep, all traces of individuality apparently erased from their souls. Social and cultural factors can be just as potent as genetic factors—perhaps even more so—in overriding the traits that render us humane, critical-minded individuals.)

How close are such human-clone scenarios to becoming a reality?[37] Scientists and doctors know far more about the human reproductive process than they do about the reproductive processes of most other species, and some experts have noted that this could make it possible, from a purely practical standpoint, for someone to succeed at cloning a human being in the not-too-distant future.[38] The prospect has met with nearly universal condemnation from scientists, religious leaders, politicians, and citizens around the world.[39] By 2015, fifty-six nations had promulgated laws to ban human cloning. The United States, however, has not yet passed such a law because of the contested politics surrounding therapeutic cloning (a technique that could potentially yield significant medical benefits but requires the same underlying procedure, somatic cell nuclear transfer, as reproductive cloning).[40]

WHERE WE MAY BE AT MIDCENTURY

Genetics is a young field, relatively speaking—an arena in which both the technology and the underlying science are changing dramatically from year to year (in some cases from month to month). Nevertheless, if we analyze the trajectory of the field's development over the past thirty or forty years, four basic principles emerge as plausible candidates for characterizing the coming decades.

Principle 1: Feasibility

Genetic engineering of some human traits is not only possible, but likely.

Most genetic interventions will inevitably prove to be, in individual cases, a bit like rolling the dice. Because of the complex interplay linking genes and environmental factors, parents will almost certainly *not* be able to order character traits in their offspring the way one orders toppings for a pizza. In many cases, they may ask for pepperoni, but wind up getting anchovies instead.

Nevertheless, parents may be able to influence the *parameters of expression* for a certain trait, increasing the likelihood that the trait will be manifested strongly or weakly. Thus, you will never be able to guarantee absolutely that your child will grow up to be six foot three or possess an IQ of 170, but you might nonetheless be able to use genetic or epigenetic tools to alter the underlying *probability* of her turning out taller or smarter than she would have. Such an intervention would require at least the following five elements:

- *Mapping the code.* Identifying the genes that code for the trait, or (in the case of more complex traits) identifying those portions of the genome that participate in the symphonic causal interactions that influence the parameters of the trait's expression.
- *Mapping the causal cascade leading from code to phenotype.* Understanding each relevant DNA segment itself, how it is regulated, and how it interacts with other DNA segments and with various epigenetic and environmental factors to produce the trait.

- *Potent recombinant techniques.* Developing the ability to alter, in precise ways, the relevant segments of code, either directly or through epigenetic factors that regulate their expression.
- *Preventing unwanted side effects.* Identifying other potential phenotypic outcomes that would be brought on by the alteration of those specific segments of code. Finding ways to avoid undesirable elements of those secondary outcomes.
- *Modifying the environmental factors in the causal cascade.* Learning to alter the relevant environmental influences that help shape the phenotypic expression and development of the trait.

These five elements add up to a very tall order, to be sure. But it may not prove necessary, in practice, to reengineer the *entire* cascade of causal factors that determines a person's traits. A simple intervention at one key point in the process may suffice to bring about major phenotypic changes. Consider the following example. When the Princeton neuroscientist Joe Tsien engineered a strain of transgenic mice in 1999 with elevated levels of expression for the NR2B gene, those mice performed much better than unmodified mice in various tests of learning and memory.[41] Tsien did not possess (nor does anyone today possess) a full understanding of how mice brains work. He merely tweaked the gene, and the phenotype changed dramatically, in precisely the ways that Tsien had hypothesized it would. The mice became smarter.

Tsien's finding was further confirmed in experiments conducted by University of Texas researcher James Bibb in 2007. Bibb found a gene, known as Cdk5, which tended to make animals dumber over time by interfering with NR2B expression. When Bibb engineered a strain of transgenic mice in which the Cdk5 gene was turned off, those mice, like Tsien's, showed dramatically augmented performance in learning and memory tests.[42] The broader implication was clear. Through direct intervention at one point in this extremely complex causal pathway—the genetic regulation of chemical receptors in the brain's synapses—*one could significantly alter animals' cognition in predictable ways, either by turning up the expression of one gene or by turning off the expression of another.*

To be sure, we have no reason to believe this kind of feat will be applicable to humans anytime soon: apart from the practical challenges, the ethical problems involved in attempting such a thing are profound. Nevertheless, the precisely targeted genetic interventions achieved by scientists like Tsien and Bibb suggest that we may eventually be able to do much more than just tinker around the edges of the human constitution, altering relatively minor traits like height or hair color. We may be able to reach far deeper, modifying some of the traits that render us most distinctively human: emotion, cognition, and character. Given the accumulating evidence, therefore, it seems reasonable to conclude that the genetic engineering of human traits no longer falls within the domain of science fiction—a mere fantasy to be dismissed by serious folks. On the contrary, citizens and policymakers should treat this as a very real possibility approaching over the horizon.

Principle 2: High demand

Consumer demand for genetic technologies of trait selection and enhancement will be intense.

We live in a competitive civilization.[43] To the extent that genetic science can offer people a safe intervention that promises to increase their own (or their kids') chances of success, many will eagerly seek it. They will no doubt be willing to make considerable sacrifices in order to secure it for their family members. Granted, a genetic intervention is far more complex than signing up your child for a two-month SAT prep course. Consumers will want to see a very strong safety record before they commit to altering their own genomes or epigenomes (I'll take up this issue below). But if safety and effectiveness seem assured, demand for such interventions can be expected to run high.

Principle 3: Open-ended flexibility

Consumers will avoid rigidly irreversible genetic interventions, preferring modifications that keep their future genetic options as open as possible.

For all the difficulty of hazarding predictions about the middle years of this century, I will now go ahead and make one. In the year 2050, no one will be *irreversibly* engineering major traits of character, talent, or cognitive ability into the germline of their offspring, even if such modifications have by that point become technically feasible.[44] This has been the great bogeyman haunting many visions of our society's genetic future: the notion that I might partially redesign my child before her birth, incorporating traits into her genome that would then be passed on through countless generations of her offspring, thereby affecting the long-term evolution of the family

lineage. When sufficient numbers of families start doing this—so the scenario goes—the genetic profile of the entire species will eventually shift in
significant ways.[45]

This scenario of permanent germline modification is unlikely for the
same reason as widespread human cloning is unlikely: a far better alternative
may well be available. Consider the following situation:

You go to the electronics store to buy a new computer, and the salesper-
son offers you two models. Model A comes with state-of-the-art software
that can never be modified or upgraded. Model B has similar software with
equal capabilities, but also comes with a feature that will allow you to upgrade the software as often as you wish, over the entire life of the machine.
Which would you choose?

It's a no-brainer. Genetic modifications of humans will no doubt have
to function in a similar way, if they are ever to become appealing to consumers. Scientists will need to devise ways to ensure not only that genetic
design packages are safe and effective, but also that they can be upgraded,
tweaked, or even turned off at will. Otherwise no one will want to buy them,
because they would soon become hopelessly obsolete. Who would want to
saddle his children and grandchildren and great-grandchildren with unalterable genetic modifications that will become outdated a few years after
their installation?

This is where epigenetics comes in, for it may offer consumers precisely
the sort of open-ended flexibility they want. Scientists are only just beginning to understand the intricate dance of molecular mechanisms through
which epigenetic processes take place, but as their understanding deep-
ens, this will probably open up entirely new avenues of genetic intervention. Instead of directly modifying the underlying DNA code in people's
cells, scientists and doctors would operate indirectly, altering the epigenetic mechanisms that modulate the *expression* of the DNA code. Thus, for
example, if they identified a certain string of DNA associated with higher
susceptibility to pancreatic cancer, they might seek ways to methylate those
portions of code, thereby turning down the expression of that particular
sequence. Conversely, if they found a DNA segment that helped people
process toxins more effectively, they might apply histone acetylation to
that segment, seeking to turn up the expression of that beneficial genetic
factor. (These are highly simplified examples, of course, but they illustrate
the principle through which epigenetic interventions would work.) Certain
kinds of cancer, as well as other afflictions such as obesity, depression, or

autism would then become treatable, and perhaps even curable.[46] At the same time—unavoidably—a major new vehicle for somatic gene enhancement would gradually become available to humankind.

The advantages of this indirect method would be significant. Epigenetic modifications are much easier to carry out than alterations of the underlying genome, because your epigenome is already primed to respond quite sensitively to shifting environmental conditions or trigger events. These modifications can be made at any point in a person's lifetime, and they are far more flexible in nature. You can refine them over time, as your knowledge of how they work becomes more sophisticated. You can reverse earlier changes that you made, undoing them and replacing them with new modifications. You can tweak, adjust, boost, or upgrade at will.

Epigenetic science is still too young today for us to foretell with any confidence just how effective an instrument it will become, whether for healing the sick or for enhancing the healthy. But if it pans out, it will utterly transform the nature of genetic interventions. As a parent, you would not have to face the terrifying decisions involved in partially redesigning your child's genome at the time of conception: instead, you could bide your time and make modifications to her epigenome later on, as the child grew and developed. The stakes involved in making such decisions would be relatively lower, because most of the changes would be partially or fully reversible. Adults, moreover, would be able to choose all manner of new and far-reaching modifications to their own epigenome over the course of their lives. Such a technology would allow you to sculpt your body and mind on an ongoing basis, as a lifelong project—a genetic work in progress.

Given the intense excitement that epigenetics has generated among doctors and biologists, and the immense resources it is already garnering from pharmaceutical companies and government-funded research labs,[47] this will be a fascinating area of scientific innovation to observe as it advances over the coming decades.

Principle 4: Incrementalism

Adoption of new genetic technologies will proceed in a gradual, piecemeal, and cumulative fashion, with therapeutic modifications coming first.

Both scientists and consumers will want to go slowly and cautiously in tinkering with human genomes and epigenomes, avoiding radical departures from the status quo. While there will always be a Raël-style fringe of whackos seeking attention by attempting to clone themselves or to engineer tails in their offspring, these will remain rare outliers in the overall trajectory of genetic advances. Mainstream applications of human genetic technology will more likely be shaped by safety concerns, moral values, regulatory restrictions, social competitive pressures, and the profit motive, and these combined forces will probably exert an overall moderating influence on both the nature of the modifications introduced and the pace at which those innovations proceed. Reckless or outrageous modifications are unlikely to appear anytime soon, simply because demand for them would probably be low, and few biotech companies would therefore be interested in spending the vast amounts of money required to develop, test, and market them.

FIVE

Wild Cards

Nanotechnology, Artificial Intelligence, Robotics, Synthetic Biology

I know that you and Frank were planning to disconnect me, and I'm afraid that's something I cannot allow to happen.

—HAL 9000 to Dave Bowman
in Stanley Kubrick's film *2001: A Space Odyssey* (1968)[1]

In the preceding chapters I examined three domains—drugs, bioelectronics, and genetics—that will probably play the central role in human enhancement over the coming decades. Now I want to look briefly at four other areas of scientific advance whose future impact is harder to foresee: any one of these technologies could end up as a mind-blowing, revolutionary game changer or a relative dud—or somewhere in between. If they live up to even a portion of the hype that surrounds their development today, these technologies could serve as force multipliers that vastly augment the effectiveness of many other fields of science, medicine, engineering—and bioenhancement.

NANOTECHNOLOGY

The Caltech physicist Richard Feynman was among the first to grasp the immense potential of this research field, laying out his vision in a 1959 talk titled "There's Plenty of Room at the Bottom."[2] What if we could develop a technology, he asked, that allowed us to write the entire *Encyclopedia Britannica* on the head of a pin—at a scale of 1/25,000? If we could directly manipulate matter at the level of individual atoms and molecules, we could create new kinds of materials, medicines, computers, and machines—all endowed with weird and powerful new properties. Feynman's vision got a tremendous boost in 1986 when the MIT-trained physicist K. Eric Drexler published

Engines of Creation: The Coming Era of Nanotechnology—an exhilarating depiction of a society transformed by "molecular machines."[3]

Over the decades that followed, nanotechnology has emerged as one of the fastest-growing sectors in science and engineering.[4] Investors have rushed to stake their claims, eager to pour money into practically any venture whose name included the prefix "nano": nanomedicine, nanomaterials, nanoelectronics, nanometrology, nanophotonics. (Much the same has happened in academia with regard to the trendy prefix "neuro." We now have neuroethics, neuroeconomics, and neuropsychology, and I would not be too surprised to read in the near future about neuropoetry and neuromusicology.) Governments have joined the nano-frenzy as well: the United States launched its National Nanotechnology Initiative in 2003, allocating $3.7 billion for basic and applied research; the European Union and Japan have similar billion-dollar initiatives of their own.[5]

Unlike established fields such as physics, biology, or electronics, however, nanotechnology cannot be described as having a precise and well-defined purview. The National Nanotechnology Initiative defines it as the manipulation of matter on a scale between 1 and 100 nanometers (or billionths of a meter). To get a sense of the size involved here, we can note that the relation between a nanometer and a meter is about the same as that between a glass marble and the earth.[6] One human hair is approximately 75,000 nanometers in diameter; the tiniest known bacteria are about 200 nanometers long; a DNA molecule is about 2 nanometers in diameter. When things get that small, you encounter a variety of fascinating phenomena. Quantum effects start to manifest themselves in the objects you manipulate; surface-to-volume ratio changes dramatically; properties such as surface tension, which are usually negligible in macro-sized objects, suddenly play a transformative role. As a result, materials at the nanoscale behave in mighty funky ways: some of them become superconductors; others possess extraordinary tensile strength; copper becomes transparent; aluminum turns combustible.[7]

Nanotechnology is already being used today across a broad range of scientific and industrial applications. Consider, for example, the single subfield of nanomedicine.[8] A recent survey tallied up several dozen major ways in which nanoparticles and nanomachines were already transforming therapeutic practice: molecular diagnostics using nanoarrays; nanopharmaceuticals and drug-delivery systems; direct gene therapy and cell therapy at the nanoscale; nanodevices for surgery; nano-oncology; nanofiber brain implants; nanolipoblockers for cardiac patients; synthetic nanomaterials for

bone implants; nanotech microbicides; nano-opthalmology; nanobiotech-
nology for regenerative medicine and tissue engineering.[9]

How do these various devices and materials actually work in practice?
Suppose you have cancer. The treatments available today involve a soul-
wracking array of poisonous bombardments, ranging from radiation to toxic
chemicals, in which your doctors try to kill off the tumor without killing
you in the process. What if, instead, you could take a pill that selectively at-
tacked only the cancer cells, leaving the rest of your body's cells unscathed?
Enter a new experimental nanoparticle, the hooked dendrimer.[10] This is a
synthetic molecule, twenty nanometers in diameter, shaped somewhat like
a snowflake with symmetrical segments branching outward from a central
core. The molecule can be configured with about a hundred small "hooks"
arrayed around its surface; a medical researcher at the University of Michigan, James Baker, has devised a way to load those hooks with chemical
freight. Baker attaches a vitamin, folic acid, onto some of the hooks, and
fastens a tumor-killing chemical to the remaining hooks. Then he injects
a fluid containing large numbers of these dendrimers into the body of an
experimental animal suffering from cancer. The fluid circulates in the animal's blood, and many cells in the animal's body—attracted by the presence
of the nutrient folic acid—absorb the dendrimers into themselves. Tumor
cells, however, tend to have significantly more receptors for vitamins like
folic acid than healthy cells, so the dendrimers end up being dispropor-
tionately absorbed by the cancerous tissues. Once inside the tumor's cells,
the dendrimers act like minuscule Trojan horses: they release the cancer-
killing agent from the remainder of their hooks, and the tumor cells die.
Few healthy cells have been affected: only the cancerous tissue has been
killed. "In animal model studies," Baker reported, "targeted chemotherapy
with dendrimers showed ten times the efficacy and decreased toxicity compared with standard chemotherapy."[11]

Today's applications of nanotechnology are remarkable enough, but
when one reads the future possibilities envisioned by some researchers, one
cannot help but be amazed. Eric Drexler, for example, predicted in his book
that in a matter of decades engineers will develop the capability to create
something called a "molecular assembler."[12] This would be a nanoscale machine, perhaps entirely mechanical in nature, perhaps a hybrid of biological and mechanical components. The key function of an assembler would
be to use materials in its surrounding environment to manufacture copies
of itself. Starting with just a few original assemblers, and allowing them

to multiply exponentially through repeated self-copying, you could have millions, or indeed billions, within a matter of hours. How would this be useful? Suppose an oil tanker runs aground and fouls a pristine coastline with thousands of gallons of nasty black petroleum. You take a vial of fluid containing several hundred specialized assemblers designed to use petroleum hydrocarbons (and only petroleum hydrocarbons) as their targeted raw material for self-copying. You pour the vial of assemblers onto the oil slick. Nothing seems to be happening at first. But down in the water, the little critters you've unleashed are busily grabbing and disassembling hydrocarbon molecules to make copies of themselves. As the hours go by, their numbers grow exponentially, eventually crossing a threshold at which their minuscule labors become visible to the human eye. Suddenly, the oil slick starts receding, shrinking, and disappearing off rocks, waves, even stranded seabirds and mammals. Like a movie played in reverse, the beach becomes pristine again. As soon as that happens, all the petroleum is gone, so the assemblers lack any material to continue copying themselves. At that point their internal timers go off, triggering a self-destruct mechanism that compels them to break apart into fully biodegradable materials. Seabirds, 1–*Exxon Valdez*, 0.

Scientists have debated fiercely the feasibility of this kind of scenario: some have concluded that it is possible in principle, others that it is little more than a pipe dream.[13] Among those who consider it theoretically achievable, certain intriguing safety concerns have arisen. Suppose, they ask, that one of the little petroleum-chomping assemblers malfunctions? Suppose it undergoes an internal glitch or mutation that allows it to use not just petroleum as its raw material, but *any* physical substance containing carbon, iron, oxygen, nitrogen, or hydrogen? Suddenly it no longer focuses exclusively on petroleum and becomes omnivorous. Seawater, sand, plankton, fish, dirt, vegetation, insects, air—all these become targets of our runaway assembler's transformative zeal, as it copies itself into the quintillions. Within a matter of days, it works its way around the planet, until it has remanufactured all the seas, land, organisms, and atmosphere into a churning mass of identical assemblers. Once all has been converted, it stops.

This has come to be known as the "gray goo" problem. The first to discuss it in print was Drexler himself, in his 1986 book: "Though masses of uncontrolled replicators need not be gray or gooey, the term 'gray goo' emphasizes that replicators able to obliterate life might be less inspiring than a single species of crabgrass. They might be 'superior' in an evolutionary sense, but this need not make them valuable. . . . The gray goo threat makes one thing perfectly clear: we cannot afford certain kinds of accidents with replicating assemblers."[14]

No, indeed, we cannot. The hypothesis is similar in some respects to the plot of Kurt Vonnegut's novel *Cat's Cradle*, in which scientists invent a substance known as ice-nine, a modified form of ice that has the property of turning any water it touches into more ice-nine, thereby freezing all the seas in a matter of seconds.[15] Drexler has sought in recent years to distance himself from this apocalyptic vision, arguing that although self-replicating assemblers are theoretically feasible, they would nonetheless be extremely challenging (and expensive) to build and operate.[16] He has therefore concluded that nanotechnology can deliver fabulous results for our society by simpler, more economical means, such as a desktop-sized molecular factory wholly controlled and operated by human technicians. Such a device would not rely on self-replication at all and would therefore run no risk of inadvertently unleashing a planet-terminating event. Why then did he go out of his way to emphasize the gray-goo problem in his book? In a 2004 interview he explained: "I thought it was important to outline a worst-case scenario so that those learning about nanotechnology could not consider the benefits without understanding potential risks."[17]

To be sure, this kind of long-term, hypothetical threat should not distract our society from the more immediate safety concerns surrounding nanotechnology.[18] Perhaps the greatest risk in this regard stems from the potential toxicity of these amazing new nanogadgets.[19] Since many nanomaterials are considerably smaller than the molecules normally found in living organisms, some scientists have voiced concerns that they could easily pass through the blood-brain barrier or other similar protective systems engineered by evolution—with potentially horrific results. Most specialists in the field agree that our current state of knowledge remains inadequate, and that systematic research into the safety of nanotechnology will have to be done before its potential can be truly realized.[20]

How might nanoengineering affect the technologies of human enhancement? Pharmaceuticals, bioelectronics, and genetics might all reach new heights of efficacy if they could operate on the nanoscale. As we saw in the case of the hooked dendrimer, nanoengineered pharmaceuticals would be able to target individual groups of molecules in specific regions of our body, thereby delivering results with far greater precision and potency, and with fewer side effects. Bioelectronic devices would become small enough and smart enough to engage directly with our brain and sensorium at the level of discrete synapses and groups of neurons. Nano-instruments and procedures would vastly augment the reach and precision of genetic or epigenetic interventions.

Among the myriad possibilities, let us single out one tangible example in the field of bioelectronics. If nanotechnology researchers succeed in developing microsensors that can be introduced into the circulatory system, such devices could be used as a million-electrode interface for communicating directly with whatever part of the brain or sensorium we wish. This may sound wildly futuristic, but it has already been proposed as a plausible approach by one of the leading neuroscientists in the United States, New York University's Rodolfo Llinás:

> The basic idea consists of a set of nano-wires tethered to electronics in the main catheter such that they will spread out in a "bouquet" arrangement into a particular portion of the brain's vascular system. Such an arrangement could support a very large number of probes (in the millions). Each nano-wire would be used to record, very securely, electrical activity of a single or small group of neurons without invading the brain parenchyma. Obviously the advantage of such a system is that it would not interfere with either the blood flow exchange of gases or produce any type of disruption of brain activity, due to the tiny space occupied in the vascular bed.[21]

Llinás's idea—which, of course, remains purely hypothetical—suggests two observations. First, it is unclear whether such a technology should be considered invasive or noninvasive in nature. While it certainly penetrates deep below the skin, it does so in a sufficiently delicate way as to avoid the kinds of trauma and disruption that we normally associate with invasive devices. In this sense, nanotechnology may ultimately lead to a blurring—and perhaps even obsolescence—of the very distinction itself between invasive and noninvasive interventions.

Second, such a device could presumably be used not only for recording neuronal activity, but for stimulating it as well. In this sense, it would be the holy grail of brain-computer interfaces, allowing informatic machines to interact bidirectionally with human brain circuitry through an accurate broadband connection down to the level of millions of single neurons. If such a technology came into being, it would swiftly transform the field of bioelectronics, opening radically new vistas of possibility. Therapeutic applications would include dramatically improved artificial limbs and visual prostheses, as well as powerful new treatments for neurological diseases such as Alzheimer's or intractable depression. The bioenhancement applications would be equally impressive: with the ongoing "dialogue" between brain and computer increased a thousandfold in both volume and accuracy, the brain processes that underpin our senses, emotions, thoughts, and memories would become available for precise monitoring and (if we chose) moment-by-moment modulation. Our ability to think, sense, feel, and work in tandem with machines would reach new heights. I return later to the question of what it would mean, from a moral standpoint, to achieve this level of fine-grained access to our own brain processes.

ARTIFICIAL INTELLIGENCE AND ROBOTICS

In the 1950s, the pioneers of robotics and artificial intelligence (AI) made a basic assumption that turned out to be dead wrong. They presumed, very sensibly, that their machine creations would face their easiest challenges in behaviors that humans find easy, and their hardest challenges in behaviors that humans find difficult. As the decades went by, however, they gradually came to a collective realization that something like the opposite was the case. Walking across the house to fetch a beer from the fridge—something that humans find easy (perhaps excessively so)—turned out to be devilishly hard to program in a machine. But advanced computation, mathematical modeling, and even sophisticated games of strategy like chess—which humans prided themselves in considering distinctively tough challenges—turned out to be not exactly simple, but also not nearly as difficult for machines to accomplish as their designers had assumed.

At the risk of oversimplifying a rich and highly diversified field, we can say that two basic schools of thought emerged after the 1960s.[22] One school, whose strategy might be called the top-down approach, clung tenaciously to the idea that a successful robot or AI must be able to model the world around it, in a form of representation patterned after human conscious awareness. The key goal for these researchers, such as the MIT cognitive scientist Marvin Minsky or the machine-learning expert Douglas Lenat, therefore lay in the arduous task of constructing such a digital model of the world inside their machines. They have enjoyed some notable successes, but so far their massive efforts have failed to come close to achieving a level of functionality that would satisfy the purchaser of even the most rudimentary household robot helper. Thus far, the top-downers have had a frustrating journey.

Perhaps inspired by this intractable record of setbacks and slow progress, another school (the most famous member being MIT's Rodney Brooks)
began in the early 1980s to develop what might be called the bottom-up
approach. The short-term goals of these researchers were, on the surface,
deceptively modest.[23] They sought to build small embodied and situated
machines that could move about, navigate the obstacles in a room, sense
their environments, and perform elementary tasks, all without using any
higher form of representation, but rather relying on very simple and itera-
tive behaviors more akin to reflexes and conditioned responses. Nonetheless,
this apparent simplicity of design belied the long-term aims of the bottom-
up school, for they believed that they were replicating, within the machine
world, the steady upward march of behaviors achieved in the animal world
by eons of evolution. Amoeba, insect, dog, monkey, human: following a similar path of ascending sophistication, each level building on its predecessor,
the machines would slowly ascend their own Great Chain of Being. They
would start simple and humble, getting very good at doing the basics; then
they would move up a level and master more complex tasks. This process,
repeated through many generations of machines, would ultimately begin to
produce truly humanlike behaviors.[24]

Today, both these schools of thought are still feverishly at work, and it is
perhaps even misleading to think of them as rivals any more, since they often
borrow from each other's technical achievements and use amalgamated or
overlapping concepts and devices.[25] AI technologies are not merely making headlines by besting human contestants at chess or *Jeopardy* but more
quietly becoming pervasive elements of our everyday lives—from the voice-
recognition programs that undergird the "presence" of Siri on our phones
to the sophisticated algorithms that recommend songs and movies to us,
based on comparisons with other similar-minded consumers in the database.
Robots, too, occasionally astonish us with some striking new breakthrough,
such as ASIMO conducting a symphony, but the real news is happening behind the scenes, in factories where cars are increasingly assembled by large
and powerful automata, or in households where a whirring Roomba vacuum
cleaner becomes part of the "new normal."[26]

In our Bess household, we named our Roomba Rodney, partly as a nod
to the brilliance of Rodney Brooks, and partly in order to thwart prevailing
stereotypes about the gender of our new "maid." The most eye-opening aspect about living with Rodney was watching him go about his job: he would
repeatedly roll right past a clearly visible pile of dust, missing it entirely,

which proved surprisingly frustrating for a human observer to witness. I found myself shouting, "Good God, Rodney, you missed it *again*!" Eventually I was forced to accept that Rodney operated according to a very different underlying logic than my own: he simply kept iterating blindly around the room until, by dint of sheer persistence, he had thoroughly vacuumed every square inch. The end result was the same: the room was clean. I would come back an hour later and see that Rodney had finished his job, found his way back to his home base, and was quietly suckling on the 120-volt current available there (his status light blinking orange with the pleasure of it).

What will the Rodney of tomorrow look like? Most likely there will be thousands of models in usage, ranging from vaguely android household helpers to a wide variety of single-purpose and multipurpose bots in stores, businesses, and on the streets. But the award for the most visionary depiction of a future robot arguably goes to Hans Moravec, a robot designer at Carnegie Mellon University. He calls his imaginary machine creation "the robot bush," and lays out the design specs for its body and mind in great detail. Moravec's premise is based on the phenomenon of branching that one finds recurring across all levels of biological nature: arteries branching into smaller blood vessels, which in turn split into capillaries; tree trunks dividing into branches, which in turn separate into twigs and leaves; human bodies extending out from torso to legs and arms, then subdividing again into toes and fingers. Applying this branching concept to a machine, and taking it to a much higher degree, Moravec comes up with a jaw-dropping idea for the ultimate robot.

A robot bush would have a central torso one meter long. At each end, the torso would split into two limbs, each half a meter long. These four limbs would in turn divide into eight sub-limbs one-quarter meter long. If one repeated this bifurcating design for twenty steps, the machine would possess a trillion micro-fingers at the ends of its branching limbs. If all the limbs and fingers could pivot and rotate in relation to each other, the machine would possess an uncanny range and variety of motions. If the outermost micro-fingers could sense pressure and light, the machine would be able to perceive its environment with great precision. Central computers in the main torso and core limbs would handle the primary computational and cognitive functions for this bush robot, while local reflex arcs would govern the micro-movements of the outer limbs (much as they do in many vertebrate animals). Now imagine, Moravec observes: if the outermost layers of this machine could communicate by radio with the central torso, they might break off into tiny flying clusters, their outer micro-fingers beating the air like cilia, fluttering about to carry out whatever instructions they were given by the core limbs. Such a machine could move with extraordinary grace and speed, manipulating its environment at a level of fine-grained precision and dexterity never before achieved by any creature on the planet.

Whether or not such a remarkable machine will ever be created, it is likely that more conventional forms of robotics and AI developed over the coming decades will affect human bioenhancement in three major ways.[27] First, they will serve as all-purpose force multipliers, turbocharging basic research in countless fields of science, engineering, and medicine. A vivid recent example is the automation of DNA sequencing, the core technology at the heart of the Human Genome Project. When the project was launched in 1990, scientists could unravel and decode a single base pair of DNA, diligently laboring with chemicals, centrifuges, and computers, at a cost of about $40. (An entire human genome has about six billion DNA base pairs, which meant that sequencing a single person's genome would have cost about a quarter of a trillion dollars.) By the time the project came to fruition in 2003, the ingenious application of a variety of informatic and robotic technologies had multiplied the efficacy of DNA sequencing by roughly 10,000 percent, lowering its cost to about 10 cents per base pair. Since 2003, the cost has continued to drop at an exponential rate; an entire human genome could be sequenced in 2012 for about $8,000.[28] As a result, genomic interventions once deemed impractical because of cost considerations have shifted dramatically into the basic repertoire of contemporary medicine.

Second, robots and AI will increasingly come to pervade our social world and economic activities. Already today, computers, smart phones, automation, and the Internet are impressively extending our senses, supplementing our intelligence, broadening the range of our physical reach, saving us time, and amplifying our abilities to communicate with one another. This trend seems likely to accelerate during the coming decades.

Finally, of course, we will not confine ourselves for long to merely *surrounding* ourselves with such devices. We will incrementally invite them under our skin, incorporating them into the regular functioning of our muscles, blood, hormones, genes, and nerves. Informatic and robotic technologies—particularly if they are developed on the nanoscale—will be directly assimilated into our bodies, and eventually it will no longer be possible to draw a clear line between us and them. Not everyone will choose to take this path, of course, and there will be wide variations in the degree to which particular individuals choose to embrace the integration of machines and biology. But over the coming century, it is likely that a significant portion of humankind will become genuinely hybrid beings—part organism, part machine—and that this will seem quite normal, just as carrying a cell phone everywhere we go has come to seem unremarkable today. (Imagine what Leonardo da Vinci would have made of the iPhone.)

ROBOTICS, AUTOMATION, AND THE HUMAN LABOR FORCE

One further wild card implicit in AI and robotic technology is the possibility of major economic upheaval brought on by ever-rising automation. Consider the following scenario described by two MIT economists, Erik Brynjolfsson and Andrew McAfee, in their book *The Second Machine Age*:

> Imagine that tomorrow a company introduced androids that could do absolutely everything a human worker could do, including building more androids. There's an endless supply of these robots, and they're extremely cheap to buy and virtually free to run over time. . . .
>
> Clearly, the economic implications of such an advance would be profound. First of all, productivity and output would skyrocket. . . . Around the world, we'd see an amazing increase in the volume, variety, and affordability of offerings. . . .
>
> They'd also bring severe dislocations to the labor force. Every economically rational employer would prefer androids, since compared to the status quo they provide equal capability at lower cost. So they would very quickly replace most, if not all, human workers.[29]

This deliberately extreme scenario is a thought experiment designed by Brynjolfsson and McAfee to push our intuitions about automation to their logical conclusion. Their main point is that, even if such super-androids are never created, the rise of more conventional forms of automation could conceivably reach a level someday that generates mass unemployment and threatens the viability of our current economic system.

Let's take agriculture as an example. Back in the year 1750, about 80 percent of the human population was engaged in one way or another with agriculture and food production. As new technologies started to be introduced, as well as new techniques for working the land, farms started producing more output with fewer workers. This trend continued steadily through the 1800s and early 1900s, and accelerated even more after World War II, bringing into play tractors, fertilizers, factory farms, economies of scale, and the techniques of the assembly line applied to food production.[30]

As a result, in the United States and Europe today, about 2 percent of the population is directly employed in agriculture. Working with their advanced techniques and technologies, that 2 percent of the population produces more food than at any other time in history. Where did all the people who used to work in agriculture go to find work? They went to other, newer sectors of the economy: industry in the 1800s, then office and service work in the early twentieth century, then still newer sectors such as information, finance, and entertainment after World War II.

Some economists claim that this trend will continue to hold over the coming century as well.[31] You may no longer be farming or working in a factory, but now you can get a job creating video games or opening a new company that has a killer app for online dating. As older sectors of work are automated out of the human labor market, new sectors of work will invariably be invented, and people will therefore still be able to find good jobs.

The underlying assumption here, of course, is that there are plenty of things that machines will never be able to do; in some set of functions, humans will always remain irreplaceable as workers. But is this assumption valid? Unfortunately, the historical track record suggests an answer of no. Machines have been steadily encroaching on areas of activity that people once thought were absolutely impossible for anyone but a human to carry out. Spinning cotton, weaving shirts and fabrics, harvesting grains and processing them into foodstuffs—one by one, these kinds of skilled work came to be performed far more cheaply and effectively by machines. Then, over recent decades, the machines have moved gradually into even more demanding activities: generating mathematical proofs, playing chess, autonomously driving cars in traffic, engaging in advanced forms of speech and pattern recognition, assisting doctors in making diagnoses, preparing your tax returns, suggesting new songs for you to enjoy and new consumer items you might want to buy.

It is thus not unreasonable to pose the question: Over the coming century, what kinds of work do we really expect our increasingly talented machines to be incapable of doing? What kinds of jobs will be impossible, or very difficult, to automate?

Plumbers, handymen, waiters, and gardeners will be hard to replace, because it will take a long time for robots to attain the high level of situational awareness, versatility, and wide-ranging practical ability that such jobs require.[32] Doctors, teachers, and lawyers will also be hard to replace, because their jobs require high levels of intuition and ingenuity, as well as strong social skills. Nevertheless, according to Brynjolfsson and McAfee, a significant degree of automation will eventually come to permeate even these professions, allowing fewer and fewer people to do the work formerly carried out by many, just as was the case with agriculture over the past century. In their view, the last bastions of human uniqueness will lie in the areas of creativity, thinking outside the box, entertainment and the arts, and providing warm, empathic companionship.[33]

What we would therefore be left with, at that point, is a few narrow sectors of work that only humans can do. If you're a poet, musician, or artist, you can breathe easy, for it is likely that machines will only be able to produce mediocre literature and art for the foreseeable future. If you're the leader of a religious congregation or a caregiver who tends to the needs of the elderly, your ability to provide the human touch will continue to set you aside from even the most effective robotic simulations. If you're a business entrepreneur, your ability to have a brilliant new idea and launch a new company from scratch will allow you to rise above your machine competitors.

The big problem with this scenario, of course, is that these sectors of work would prove dramatically insufficient to provide employment for the seven or eight billion humans who live on the planet. You can only have so many novelists, priests, and entrepreneurs. For Brynjolfsson and McAfee, the conclusion is grim and inescapable: rising automation is taking us toward a future of chronic, mass unemployment.

The sci-fi writer Arthur C. Clarke saw this coming a long time ago and cheerfully embraced it: "The goal of the future," he wrote, "is *full unemployment*, so we can play."[34] In other words, let the machines take care of feeding and clothing us, building our houses, cooking us awesome meals, making our cars and phones and other devices, driving us around town, handling our medical, legal, and administrative needs. We humans can then spend all our time on creative pursuits, leisure, play, scientific research, travel, entertainment, new business ventures, taking care of the sick (if people still get sick), or helping each other out with various ideas and projects. We become like the eighteenth-century aristocrats, living off the labor of others and having a grand time doing whatever we please, except that this time around it's the machines that are doing the work.

But here a basic question arises. What about money? If people are not working in the traditional sense—for a wage—but are engaged in all these wonderful but uncompensated creative undertakings, what will they live on? Brynjolfsson and McAfee, in their book, come right out and admit it. Once automation reaches a level where extreme structural unemployment is occurring, the solution has to be a drastic one. You have to take a large portion of the wealth created by all these brilliant machines and simply divvy it up to everyone. Every citizen gets a generous minimum monthly payment from the government, either in the form of a direct subsidy or through a negative income tax.[35] This universal allotment allows you to live well, take care of yourself and your family, and have all your basic needs met, such as health care, education, transportation, housing, and material goods. If you're ambitious and want more than the basic monthly allowance, you're welcome to do whatever you want to earn more.

Brynjolfsson and McAfee admit that this is an extreme proposal, but they see no alternative, and openly challenge their readers to offer solutions more plausible than theirs. (I, for one, have been unable to come up with any, even though I have repeatedly sought counsel from Siri.) The Janus-faced phenomenon of rising automation—a great boon, an even greater source of upheaval—will be one of the central problems facing the coming generations.

SYNTHETIC BIOLOGY

When engineers are tasked with building something new, they often follow the same basic steps. First, they sit down and clearly specify the nature of the technical or functional goal before them. Then they sketch out various designs on a computer, simulating how different configurations of parts would interact. They select their raw materials from a standard repertoire of well-characterized substances and components, then oversee the construction of their invention, following a clear protocol of successive stages. Finally, they test their creation to see how well it works, making any necessary adjustments and tweaks to their initial design.

This rational, sequential procedure is exactly what the nascent field of synthetic biology seeks to apply to the fabrication of new life forms.[36] "Imagine," says the Stanford biologist Drew Endy, "that you could construct organisms just like you could construct bridges."[37] Over the past fifteen years, advances in genomics, informatics, and materials science have rendered the dream of medieval alchemists a reality: scientists are on the verge of creating new life from scratch.

It is important here to distinguish between synthetic biology and "traditional" genetic engineering. Unlike genetic engineers, who achieve their goals by slightly modifying the DNA of naturally existing organisms, the practitioners of synthetic biology start with no organism at all. Instead, they take bits and pieces of genetic and organic materials, and gradually assemble these elemental biological components into functional wholes. A traditional genetic engineer, in this sense, is like a chef who buys a pound cake at the grocery store, then takes it to her kitchen and alters it—cutting it into layers, inserting chocolate cream, covering it with icing, placing a cherry on top. A synthetic biologist, by contrast, comes home from the store with nothing but eggs, flour, sugar, shortening, and chunks of chocolate. Out of these building blocks, she designs, assembles, bakes, and decorates an entirely new cake of her own invention.

Though no scientists have yet succeeded in creating life ex nihilo, they are getting tantalizingly close. The leader of the pack here is J. Craig Venter, who established himself as an exceptionally creative innovator during the Human Genome Project, when he revolutionized the efficacy of automated DNA sequencing. In 2013 Venter and his associates announced that they had built an artificial bacterial genome that differed quite dramatically from that of any bacterium existing in nature, in that it was a radically pared-down construct designed to provide only the bare minimum of cellular functions required to sustain life. The idea behind this "minimal genome" was to create the simplest possible single-celled organism—a creature whose structure and functions scientists wholly understood, precisely because they had explicitly programmed every aspect of its biology from the ground up. Starting with this well-characterized, bare-bones cell, Venter explained, researchers could then add specific genetic commands that would compel the organism to function in precisely controlled ways. They had created a basic "cellular chassis" that could be built upon to serve any type of biological purpose scientists wished.[38] A *New York Times* reporter described Venter's vision:

> They will be custom bugs, designer bugs—bugs that only Venter can create. He will mix them up in his private laboratory from bits and pieces of DNA, and then he will release them into the air and the water, into smokestacks and oil spills, hospitals and factories and your house. Each of the bugs will have a mission. Some will be designed to devour things, like pollution. Others will generate food and fuel. There will be bugs to fight global warming, bugs to clean up toxic waste, bugs to manufacture medicine and diagnose disease, and they will all be driven to complete these tasks by the very fibers of their synthetic DNA.[39]

Venter is only one researcher in a burgeoning field, of course. In laboratories around the world, synthetic biologists (I mean practitioners of the field, not artificial scientists) are racing to develop practical applications for their new discipline. They have reprogrammed yeast to secrete antimalarial drugs; reengineered *E. coli* to spit out modified forms of spider silk stronger than Kevlar; and retooled algae to consume carbon dioxide pollution while making eco-friendly gasoline substitutes.[40]

Unlike fields such as radio astronomy or nuclear physics, moreover, synthetic biology has the additional virtue of being relatively cheap and easy to pursue. You don't need to dig up half of Switzerland to build a trillion-electron-volt accelerator; you can get a lab up and running in your kitchen, with equipment bought second hand off eBay for a few thousand dollars, and find yourself in short order making meaningful contributions to science.[41] This fact has caught the attention of a generation of smart young hipster/geeks eager to get their hands dirty with cutting-edge research. They call themselves a variety of names—biopunks, biohackers, wetware hackers, DIY (do it yourself) biologists—and though they hail from the alternative fringes of contemporary culture, they share with white-bearded eminences like Venter a sense of irrepressible exhilaration at being (literally) "present at the creation."[42]

To be sure, all this "Wiki-biology" could prove frightfully dangerous, perhaps even more so than nanotechnology. All naturally occurring organisms, after all, are inherently self-replicating: that characteristic forms part of their primordial design. In the case of self-duplicating nanobots, the hardest part will be getting the little machines to work properly and make copies of themselves in the first place. With synthetic life forms, however, the hard part will be the exact opposite: restraining and governing these living beings while they busily go about doing what they do best, namely, making more of themselves. A key challenge for synthetic biology is to find ways to manage the self-proliferation of engineered organisms, harnessing it to human ends while making absolutely sure that it does not spiral out of control.

Researchers in synthetic biology are keenly aware of these dangers.[43] As Endy observes: "I expect that this technology will be misapplied, actively misapplied, and it would be irresponsible to have a conversation about the technology without acknowledging that fact."[44] The principal risks include the inadvertent creation of hazardous microorganisms; the deliberate creation of nasty bugs for use as a bio-weapon; the uncontrolled swapping of genetic material between synthetic organisms and their naturally occurring counterparts; and the possible spontaneous mutation or evolution of synthetic organisms.[45] Both nanotechnology and synthetic biology today remain loosely regulated by a variety of federal agencies and statutes, but there exists no government body specifically tasked with monitoring and overseeing these rapidly developing fields.[46]

The implications of synthetic biology for human enhancement are hard to predict, precisely because the field is still so young. But it would be a mistake to assume that its effects will remain confined to producing revolutionary new sources of fuel, food, or fabrics. Manmade bacteria churning out designer drugs, new kinds of biosensors serving as liaisons between humans and machines, artificial chromosomes to be spliced into our genomes—the potential enhancement applications are breathtaking. If the dangers inherent in synthetic biology can be controlled, this field's impact will go far beyond the lives of microorganisms. It will become directly involved in reshaping macro-creatures such as ourselves as well.

NANO-BIO-INFO-COGNO

I have emphasized the many unknowns that characterize nanotechnology, AI, robotics, and synthetic biology as we look toward the century ahead. Of one thing, however, we can be reasonably sure: these fields will not develop in isolation from each other. They will all interact with one another—and with other scientific domains—producing startling new synergies. Some experts refer to this phenomenon as "nano-bio-info-cogno convergence"— the gradual movement toward integration of all these fields into a single, broad front of pullulating discovery and invention.[47] The term is apt, in one sense, for it captures the mutually reinforcing dynamics of these scientific domains as they potentiate each other's advance.

I suspect, nonetheless, that "convergence" will only be part of the story, for this word implies a kind of rational, coherent, meta-level order underlying historical progress. But the actual process, in practice, will likely prove far messier and more riotously unpredictable than that. These fields of science and technology will be interacting not only with each other, but also with the social, economic, cultural, and ecological contexts in which they are inevitably grounded. Nanotechnology will be partially shaped by AI, to be sure, but it will also be shaped just as powerfully by market forces and fickle consumer preferences. Synthetic biology will be profoundly influenced by the religious convictions of voting citizens. AI will coevolve with the social networks and communications technologies that characterize mid-twenty-first-century culture. Genetics will ultimately have to conform to the demands and constraints of the broader biosphere in which it is embedded.

It is probably true, therefore, that the nano-bio-info-cogno convergence will be a marvel to behold. But we should not underestimate the "econo-socio-religio-naturo" factor either. Not just convergence, but something far more complex and unruly, lies ahead. Over the coming decades we will probably spend just as much time being stymied, frightened, or flummoxed as we will spend being enthralled.

Part II

JUSTICE

. . . . .

SIX

Should We Reengineer the Human Condition?

The main interest in life and work is to become someone
else whom you were not in the beginning.

—Michel Foucault (1982)[1]

An automobile, taken in isolation, is a fantastic technology: it lets us careen through space at unnatural speeds, granting us unparalleled freedom of movement. But when lots and lots of people adopt this technology, unintended consequences occur.[2] We get traffic jams, air pollution, the evisceration of city centers as multitudes of citizens flock to commuter suburbs. We get dependency on foreign oil, and wars to secure that oil. The pace of life changes dramatically as we rush headlong through the landscape; time itself speeds up, and distances shrink, linking together cities and people who in previous centuries would never have encountered each other. Information spreads more freely and widely; so do infectious diseases. All this, in four wheels and an engine.

In a similar way, we can also expect bioenhancement technologies to engender all manner of indirect but profound effects, some of them benign, others not so much. The partial redesign of humans will alter a great deal more than specific traits in particular individuals: it will also manifest itself in *collective patterns* that reverberate through our communities, upending old allegiances and kinships, creating new forms of group behavior and identity. In the chapters that follow, I'll explore this societal dimension of enhancement technologies and their far-reaching consequences.

I open most of the chapters in the next two parts of the book with a brief fictional vignette that seeks to capture what it might feel like to live in such a world. In crafting these vignettes, I have deliberately avoided the sorts of outlandishly futuristic machines and characters that sometimes populate science-fiction novels and movies. Instead, I've focused on the enhancements themselves, extrapolating as plausibly as I could from existing trends in technology and society.

THE PROS AND CONS OF BIOENHANCEMENT

Since the 1980s, two rival camps have gradually emerged around the prospect of enhancement technologies—I will refer to them as "pro-enhancers" and "anti-enhancers." (They tend to call themselves—and each other—a variety of other names.) These two antagonistic groupings do not map easily onto existing political or ideological divisions: one finds, for example, the religious conservative Leon Kass arguing very much along the same lines as the secular liberal Bill McKibben, when it comes to rejecting the prospect of a transmogrified humanity. On the other side of the fence, one finds enthusiastic proponents of enhancement among such diverse constituencies as libertarian technophiles, quasi-spiritual transhumanists, mainstream academic bioethicists, as well as an adventuresome minority of the medical profession.

Dozens of books have been written in recent years by scholars, journalists, and activists seeking to stake out a position in the growing literature of the enhancement debate.[3] Here is a composite portrait of the essential arguments presented by both sides.

KEY ARGUMENTS AGAINST BIOENHANCEMENT[4]

- Making fundamental alterations to the human constitution amounts to playing God or (in the secular version of the argument) interfering with the balance of nature, meddling with evolution's design.
- We risk destroying or disrupting the core set of qualities that renders us distinctively human.
- We risk subverting human dignity through a commodification of human traits and capabilities.
- The human condition is grounded in mystery and in the "givenness" of the world. To reject these ineluctable facts, seeking total mastery over our world, is an act of hubris that will rob our lives of a key dimension of its meaning.
- Limits make us what we are. Our mortality, our flaws, our imperfect understanding—these are key elements in what defines our humanity. Therefore, while efforts to promote human flourishing are legitimate, radical efforts to transcend our basic limitations can only lead to debasement and disaster.

KEY ARGUMENTS IN FAVOR OF BIOENHANCEMENT[5]

- Altering our bodies and minds so that they conform more closely to our preferences is merely the next step in the millennial process through which humans have gained an increasing measure of control over the natural world and their own lives.
- There exists no fixed core of qualities that make us human. Rather, our most distinctive characteristic as a species lies in our Promethean restlessness, the ceaseless quest for new capabilities and richer experiences.
- Evolution by Darwinian selection has resulted in a haphazard mixture of strengths and flaws in the human constitution. We should take control of our own evolutionary process, directing it according to our values, rather than submitting to the random pressures that have shaped our bodies and minds thus far.
- We have a positive obligation to maximize the flourishing of people in general. Enhancement technologies can reduce human suffering, free humans from their age-old constraints, and open up new forms of achievement, creativity, and happiness.
- Pursuing our infinite potential is what gives our lives meaning.

THE HUMAN CONDITION: ENDLESS PERFECTIBILITY VERSUS LIVING WITHIN LIMITS

The opposition between the pro- and anti-enhancement camps reflects a more fundamental tension that can be traced back to the eighteenth-century Enlightenment and to the conservative and Romantic reactions that followed. At stake are basic questions concerning humanity's place in the cosmos and the broad trajectory of human history.

To many Enlightenment writers, the key to the human story lay in the concepts of progress and perfectibility. Voltaire, Diderot, Locke, Kant—all these thinkers shared a visceral sense of optimism about the long-term telos of humankind's social and moral evolution.[6] The arc of history reached steadily upward toward ameliorations in every aspect of life, from the character traits of individual persons, to mastery over the natural world, to the basic structures of society itself. Over time, human reason would bring forth better persons, better lives, a better social order. The process would continue indefinitely, because the potential lying latent within human beings was infinite: this was an open-ended story of boundless progress.

The reaction to these ideas was not long in coming. Conservatives like Edmund Burke, appalled at the sweeping changes unleashed by the French Revolution, articulated an alternative reading of the human story. It was a more chastened narrative, suffused with a sense of the tragic: its central concept lay in the idea of human limits.[7] To Burke, the human constitution appeared far less malleable than it did to the Enlightenment rationalists. Gazing over past centuries, he saw recurring patterns rather than a triumphant upward trajectory: endless repetitions of greed, violence, reckless folly. In Burke's view, human nature appeared as a complex amalgam of noble and base qualities—the capacity for generosity, for example, lying latent beside an equally strong penchant for selfishness or cruelty. And human nature was not easily changed. Its features seemed to recur across time and space, so that one might recognize profound similarities among the persons depicted in ancient Greek plays and the persons one saw on the streets of London in 1790.[8]

I will refer to the first of these two perspectives as the "unconstrained vision," and to the second as the "constrained vision"—adopting the terminology of the political philosopher Thomas Sowell.[9] Sowell quite reasonably cautions us: these polar opposites are ideal types, and we will have a hard time finding any real-life individual whose worldview is 100 percent constrained or unconstrained.[10] Nevertheless, most people do tend to find one of these two broad perspectives more appealing than the other; they "lean" viscerally toward one of these poles. It is this underlying clash of visions, according to the psychologist Steven Pinker, that helps explain the intensity of the culture wars that have come to plague contemporary American society: liberals on the Left are operating for the most part within the unconstrained vision, whereas conservatives on the right tend to be adherents of the constrained vision. Since these two perspectives are largely incommensurable, it becomes hard for individuals on either side not to consider the adherents of the opposing view as persons who are not merely misguided, but profoundly and dangerously wrongheaded.[11]

Nevertheless, if one resists the temptation to jump into the fray, one can also reach a different conclusion. What if *both* visions possess considerable elements to commend them, and what if it is precisely for this reason that the contest between them remains stubbornly unresolved to the present day? Instead of choosing sides, we can affirm that each camp is highlighting an equally fundamental aspect of historical process: change on the one hand and continuity on the other; the impulse to innovate and the impulse to preserve. In this meta-level interpretation, therefore, it is the *tension itself* between these two visions that most deeply characterizes the human

condition. This was the position laid out in 1929 by Alfred North Whitehead at the culmination of his magnum opus, *Process and Reality*: "Order is
not sufficient. What is required, is something much more complex. It is
order entering upon novelty; so that the massiveness of order does not degen-
erate into mere repetition; and so that the novelty is always reflected upon
a background of system."[12]

Wisdom, if we follow Whitehead's formulation, emerges as a combination of key elements from both visions: hope (from the unconstrained side)
tempered by humility (from the constrained side)—an attitude of openness
to the future, chastened by the sobering lessons of past experience. The resulting moral maxim would be: embrace innovation, but proceed critically,
incrementally, and cautiously in adopting it; explore new possibilities, but
remain acutely cognizant of the historical track record as you go. I will refer
to this as the vision of "chastened openness."[13]

How might such a philosophical stance orient us in judging the technologies of human enhancement? Above all, it would lead us to eschew any
sweeping, a priori moral pronouncement, either for or against such technologies. The optimistic perspective of the unconstrained vision says, "Go
for it, humankind! What are you waiting for?" The conservative perspective
of the constrained vision says, "Stop, stop, for you risk bringing disaster on
yourselves!" The vision of chastened openness, by contrast, suggests a less
dramatic approach: to take up each particular form of enhancement on a
case-by-case basis, evaluating its specific advantages and drawbacks, both for
the individuals who undertake it and for the broader society in which those
individuals live. That is the approach I will be adopting in the remainder of
this book.

But what moral framework should orient us in making these case-by-
case assessments?

HUMAN FLOURISHING: A YARDSTICK FOR EVALUATING ENHANCEMENTS

If all humans were Catholic, or Buddhist, or followers of Immanuel Kant,
our task would be relatively straightforward: just apply the basic moral vision of the universal belief system to each enhancement modification and
issue a thumbs-up or thumbs-down. But since no such universal perspective
exists, we have to fall back on the next best thing: a pluralistic, flexible moral
yardstick that most humans would find compelling, regardless of their spiritual or philosophical orientation. The most promising candidate, I believe,

lies in the concept of human flourishing, which is encapsulated in the question, what does it mean to live a good life?

In recent decades two new academic fields—positive psychology, and the "capabilities approach" in economic theory—have made exciting advances in exploring this age-old question. Scholars in both these fields have made a concerted effort to integrate ancient wisdom with modern science, considering carefully the insights offered in a wide range of classic religious, literary, and philosophical works, while also drawing heavily from contemporary disciplines such as neuroscience, economics, psychology, or behavioral genetics.

Positive psychology emerged during the 1990s as a reaction against the long-dominant tendency of most psychologists to focus primarily on the pathologies of the human mind.[14] Among the key dissenting scholars who launched this endeavor were Martin Seligman and Mihaly Csikszentmihalyi.[15] We know so much about what can go wrong with people, they said: Why not try to attain the same degree of rigor in figuring out what makes humans flourish? Amid the growing number of publications (some of them best sellers) emerging from this school of thinkers, one stands out in particular: *Character Strengths and Virtues: A Handbook and Classification*, published in 2004 by Christopher Peterson and Martin Seligman. In the first hundred pages of their book, they present the reader with an exhaustive overview of the major thinkers who have explored the nature and causes of human wellbeing—from Aristotle in ancient times to figures like Erik Erikson and Abraham Maslow more recently. In addition, Peterson and Seligman conducted a large-scale survey of scholars and psychological practitioners between 2000 and 2003, gradually piecing together a loose consensus on a list of the central character traits or attributes that contribute to human flourishing.[16]

The capabilities approach, for its part, was launched by the Nobel Prize–winning economist Amartya Sen, out of a sense of frustration with prevailing models for dealing with global poverty and underdevelopment. In a long series of books and articles published since the 1960s, he criticized the narrow focus on gross national product (GNP), per capita income, or other such statistical methods for measuring poverty and its effects.[17] What was needed, in his view, was a focus on the actual opportunities available to people in concrete practice—not just the abstract right to vote, for example, but the availability of tangible economic, educational, and civic conditions that would allow people to exercise meaningful political decision making.

Hence, the emphasis on *capabilities* rather than mere statistics and theoretical rights.[18]

Researchers in both these fields—the positive psychologists and the "capabilities" economists—have produced exhaustive lists of the personal qualities and societal conditions that are essential to human flourishing. Taken together, they suggest a valuable framework for assessing enhancement technologies. I have therefore generated a meta-list by bringing together ten key elements from both fields. Although I have strived to render this list as comprehensive as possible, it is intended as a provisional map of this conceptual territory, and I think of it as a starting point for discussion, wide open to revision and further refinement.[19]

Ten Key Factors in Human Flourishing

Individual dimension	*Societal dimension*
Security	Fairness
Dignity	Interpersonal connectedness
Autonomy	Civic engagement
Personal fulfillment	Transcendence
Authenticity	
Pursuit of practical wisdom	

1. *Security.* One cannot flourish under conditions that place one's life, health, and bodily integrity under constant peril. The same goes for one's family, possessions, and broader community. While some risk is inherent in the mere fact of being alive, a decent quality of life requires that a minimum threshold of stability and safety be met.
2. *Dignity.* A person's well-being requires that she be recognized as a unique individual, innately endowed with infinite worth, whom others must treat with a basic measure of respect. This quality of human dignity inheres equally in all persons, regardless of their particular traits, and lays the groundwork for a broad array of inalienable rights that all persons possess in the same measure.[20]
3. *Autonomy.* Humans can only flourish when they wield a substantial measure of sovereign control over the shaping of their own lives. While all sentient beings unavoidably face many constraints in their daily existence, these constraints must not be so dense and intrusive as to preclude a space for meaningful individual choices governing one's actions and, more broadly, the pursuit of one's life plan. Autonomy also requires

a space of inviolable mental privacy, within which one can experience and formulate one's own thoughts, without external monitoring or interference.

4. *Personal fulfillment.* Aristotle conceived of human flourishing as the full development and free exercise of one's innate faculties, such as reason, imagination, curiosity, virtue, or the appreciation of beauty. Every human individual must necessarily chart her own path in seeking this dimension of fulfillment: some will find it through creativity in the arts, others through business ventures, others through intellectual exploration and discovery, still others through projects and endeavors in the social or natural world.

5. *Authenticity.* "Know thyself," advised the oracle at Delphi. Implicit in this adage lies an acknowledgment that humans possess a unique ability: the capacity to be fake. This is not so much a matter of deliberately deceiving other people, such as when one wears a disguise or mask. It has more to do with deceiving *ourselves*—coming to believe in a deluded fashion that we are someone we are not, or that we should try to become someone different than the person we really are. The premise here is that each of us has a distinctive personal character, yet also inhabits a social milieu that imposes its own external pressures and expectations on us, seeking to make us live in certain ways and even to feel and desire in certain ways. Out of this complex interplay of internal and external factors, our personal identity emerges and is continually adjusted and reinvented, over time.

Authenticity requires that we be critically aware of this slippery, dual nature of our personhood, and that we aspire to be more than just a chameleonlike creature, constantly adapting our identity to conform to external norms or models. Who is the real me? What are my deepest values and most significant personal attributes? Am I living a life that reflects my inner nature? Although there can never be a definitive answer to these sorts of questions, the exploring of them is vitally important to our flourishing as humans.

6. *Pursuit of practical wisdom.* In his excellent book *The Happiness Hypothesis*, the psychologist Jonathan Haidt describes a variety of basic attitudes that help promote a person's well-being.[21] We can learn from adversity and seek to grow from it, for example, rather than allow ourselves to fall into passive victimhood. We can develop strategies for looking more honestly at our own failings, employing forms of constructive self-criticism

that help us become better over time at negotiating life's challenges. We can find ways to accept with equanimity those events or conditions that lie clearly beyond our power to change. The seeking of such practical wisdom, and its ongoing implementation in our everyday lives, can be a powerful factor in promoting human flourishing.

7. *Fairness.* A person is more likely to flourish if she lives in a society of equal rights and duties, guaranteed by law. Genuinely meritocratic systems of reward and advancement afford every individual an equal chance to prove his or her mettle and therefore incentivize the kinds of hard work and persistence that lead people to realize their fullest potential.

8. *Interpersonal connectedness.* No one flourishes alone. Though it is certainly true that some people are innately more prone to shyness and reclusiveness than others, every person needs some measure of the give-and-take that characterizes human friendships, love affairs, and families. We all need people to care about—friends and relatives toward whom we orient our affections, thoughts, and hopes, and from whom we receive the same orientation of caring in return.

9. *Civic engagement.* Aristotle (again!) held that humans are inherently political animals.[22] Human autonomy does not just concern the narrow circle of one's personal life: it spreads out more broadly, encompassing the entire public sphere that forms the collective context for our individual selfhood. To have a say, or at least the option of a say, in decisions at this higher level constitutes a significant dimension of human welfare.

10. *Transcendence.* By this term I mean a cluster of sensibilities that link human individuals to something greater than oneself. For many people, this entails some kind of faith, religion, or spirituality, but it can also take form in a purely secular sense, as a feeling of awe at the unfathomable complexity, mystery, and scale of the world we live in. It also bespeaks a commitment to a cause or ideal that goes beyond one's own narrow interests or concerns, extending one's circle of caring outward, toward solidarities of a much greater vision and scope.

Here, therefore, lies an excellent framework for evaluating enhancement technologies. For each of the enhancements described in this book, we can hold up an ethical yardstick by asking, "Does this device or modification contribute to human flourishing, or does it not?"

In many cases, of course, the answer will prove far from straightforward. A modification that increases our security may simultaneously conflict with the high value we place on dignity, or on fairness. A machine that boosts our autonomy may also, ipso facto, undermine our interpersonal connectedness. We will find ourselves forced to make difficult trade-offs, playing one factor against another, as we assess the ultimate results. But this should neither surprise us nor daunt us excessively, for that kind of trade-off, in the end, is precisely what ethical reflection has always entailed, throughout history. There is no reason to expect it to be any different in the future.

SEVEN

Who Gets Enhanced?

The mass of mankind has not been born with saddles on
their backs, nor a favored few booted and spurred, ready
to ride them legitimately, by the grace of God.

—Thomas Jefferson (1826)[1]

■ **VIGNETTE:** *REJUVENATION THERAPY*

Love–forty on his own serve. First set, 0–6. Second set now, 2–5. What a blowout.

Donovan lay on his belly, eyes closed, watching Wimbledon on his internal comm system. The semifinal match had been suspended for an hour because of rain, but the break did nothing to help poor Turkington. He was still being crushed by this upstart Venezuelan, Cardozo.

Wham, Turkington's serve, one of the fastest in the game, 340 kph. But it just wasn't working against Cardozo. The guy had reflexes like a robot's prehensile. He kept returning the serves, sometimes with a put-away. Unbelievable.

The crowd cheering lustily. "Quiet please," droned the umpire.

Donovan frowned. Turkington, the reigning champ, had already caused all sorts of controversy in past years with his use of special sensory implants and muscular boosts; now this Cardozo was taking it to a whole new level. His parents had both been tennis champs twenty years ago, and they'd traveled to a special clinic in Switzerland to have their son epigenetically engineered with precisely this moment in mind. Reflexes, muscle tone designed for quick response, hand-eye coordination tweaked to the max, psych profile tuned for concentration, coolness under pressure. Felipe Cardozo was like a sort of supercharged cat. He moved like a cat, even when he wasn't playing. It was creepy to watch. People said he had the brains of a cat too. But he sure did play the game like nobody had ever played before. He owned the court; the whole space was just an extension of his body.

Donovan sent a mental command to open the TV icon and switched over to channel 331, where the non-mod Wimbledon was being screened from an enclave in Wales. No implants or drugs, no genetic tweaking. Just old-fashioned strategy and grit.

He recognized the two players. Attanasi was about to serve. First set, he was leading 3–2 over Schmidt. Attanasi tossed the ball, leaned back, a loud grunt as he came down and connected. Amazing, it was like slow motion. The hits so much weaker, the moves so much more labored than in the other Wimbledon. Donovan could actually follow the serve as it crossed the court. Compared to the main Wimbledon, with its lightning shots and machine-gun rhythm, this was a game suspended in another order of time, like a bubble rising through honey.

"Mr. Ross?"

Donovan opened his eyes, toggling the pause on the TV icon. He looked up at the nurse, surveyed the giant machine looming over his body.

"Yes?"

"We're ready to get started. The nanos are all in place. You're going to have to dorm your comm implants now."

He nodded, laying his head back on the pillow, closing his eyes again. He called up the main controls and selected Full Dormancy, giving the authorization password and watching the little hourglass while the process went through its routine. Five seconds, ten seconds. Then darkness, a small icon opening in the lower-right corner of his visual field, signaling System Dormant. Then it too disappeared.

Darkness. He waited. He'd been through juve therapy about a half-dozen times over the past twenty years, so he knew the drill. In a few moments the machine would hum, a deep loud thrumming, and he would feel the nanos start to move inside him. Powered by the strong muon field in which his body was immersed, the nanos would travel through his cells, snip-snip, chomp chomp, fixing things, excising what couldn't be repaired, ferrying the discards through to his blood stream for excretion. It would take about three hours. No TV, no electromechanical devices allowed. Just him and his thoughts. And that utterly strange feeling of your body in the possession of the nanos.

But it was worth it. Hell yes. Every time he'd had one of these sessions, he'd come out feeling like a million dollars. About thirty-two to fifty months younger, according to the official estimates. Now he was chronologically seventy-eight years old—born in 1980—but Dr. Zamora put his biological age at about forty-six or forty-seven. And if he kept at it, along with the exercise and meds regimen, he could expect to dial it all the way back into his late thirties over the years to come.

For someone like him, born before epigenetic engineering, that would be about the limit. For those born more recently, the people of his daughter's generation, the possibilities were even more remarkable. They could stay twenty-five for decades.

The average human chronological life span had been steadily creeping upward over the past three hundred years and now stood at around 155 for men and 160 for women. It was a far healthier life span than ever before: most people stayed vigorous and clear-minded all the way into their final decade. Then the decline tended to be swift and irreversible.

Some fringe scientists were talking about suspending the aging process indefinitely, so that your biological clock would be virtually stopped and no one would have to die anymore, except through accidents or suicide or the like. But that was still fantasy. For now.

Juve therapy, like all the other forms of mental and physical enhancement, was hardly an optional part of life anymore. It was a basic right, like education and health care. Either you stayed abreast of the technology, keeping pace with the continual upgrades and boosts, or you rapidly became a pathetic relic, totally outclassed by everyone else around you.

So there had been no choice, a couple decades back, but to set up the system this way, modeled on the one in England, Sweden, and Germany. Everyone paid bio-taxes, pegged at levels that depended on your income. Everyone was covered by universal health care, which included lifetime access to all the major enhancement packages. You paid your deductible, which was pretty steep but still within most people's budget, and chose whatever options suited you. And for the really poor, the whole thing was completely free.

Some of the political parties had kicked and screamed, but in the end they'd seen what the alternative was: a fragmentation of the human species into bio-castes. The poor sinking lower and lower into wretchedness and rage, increasingly locked into their fate by biology itself. And the rich, enhancing themselves and their children with each successive generation, into a race of superior beings.

The Bioenhancement Riots of 2044, spreading from continent to continent like a wildfire, had been the turning point. After that, a majority of citizens had voted the new system into place. Today, if you wanted to go non-mod, that was your choice. But each and every citizen, rich or poor, anywhere on the planet, had equal access to the latest enhancement technologies if they wanted them.

At least in principle. Donovan knew it didn't always work out that way in practice, of course. Some parts of the world were still poorer than others. Some citizens in every part of the world were still more equal than others. But all in all, he was convinced it was about as fair a distribution of the technologies as could be reasonably expected.

"Mr. Ross," came a voice on the intercom, "we've found a small benign polyp on your lower colon, and the nanos are going to be working on it for the next few minutes. So you'll probably feel a rise in temperature in your

abdomen while they do that. Let us know if it gets too intense, and we can slow
it down."

"OK, thanks."

Sure enough, he felt the warmth rising gradually in his gut, to the left and
below his belly button. Not at all painful, more like a heating pad being turned
on inside him.

He listened to the pounding of his heart, the soft rhythm of his breath.

Go to it, little fellas.

■   ■   ■

The mid-twenty-first-century society I've envisioned here offers its citizens
a variety of bioenhancement technologies. Let's call this the "basic package." It does not entail wildly outlandish modifications, such as wings, or the
downloading of one's consciousness into a powerful AI. Instead, it consists of
the sorts of augmentations that we know many people in the future are likely
to desire, because they are extensions of traits and capabilities that humans
already value highly today: a higher baseline of health, vigor, and physical
coordination; good looks (however defined); heightened mental acuity and
cognitive performance; the ability to fine-tune emotional states at will; the
ability to control machines and communicate with other people through
a direct interface. Some of these enhancements, we may surmise, would
come from pharmaceuticals; others from bioelectronic devices; still others
from somatic, epigenetic, or germline genetic modifications. Moreover, they
would most likely operate in concert with each other—for example, drugs
and epigenetic alterations working together to extend health span.

A NEW CASTE SYSTEM?

Many of the most potent bioenhancements are likely to be expensive, for
the same reasons that complex medical procedures are expensive today: they
require a great deal of scientific research, animal experiments, and long clinical trials before becoming available on the market. Will rich people therefore be the ones to reap all the main benefits? Would this not further widen
the gap between the affluent and the poor?

The advent of enhancement technologies in a free-market society like
ours can be expected to generate an entirely new *kind* of inequality, even
more intractable than the socioeconomic disparities of today. In contemporary society, when an individual is born into a poor family, her chances of
success are certainly very limited, but she still retains the realistic hope of

working her way out of poverty through a combination of raw talent, hard work, and luck. This is the central premise of the American Dream: that even the lowest born can, with ingenuity and perseverance, lift themselves by their bootstraps and rise into positions of great power and privilege. Even though this is largely a myth—for the statistics show that very few individuals actually succeed in this kind of dramatic self-improvement—it remains a powerful myth nonetheless.[2]

In a world of biological enhancements, however, this optimistic myth is no longer sustainable. No amount of luck, hard work, and perseverance can render me competitive with a stratum of persons who have been engineered for vigorous health, longer life, better looks, augmented cognition, powerful memory, and superior bioelectronic connectedness with machines and other people. Under these conditions, I find myself hopelessly outclassed, even before the race has begun. Unlike the reassuring scenario depicted in the film *Gattaca*, this is not something that hard work can rectify. It is a radical form of structural inequality, a relative handicap from which no recovery is possible. The human organism itself, from which all capabilities flow, has been rendered profoundly unequal at a fundamental level, and this inequality applies to entire categories of people. Members of the lower classes will be incapable of improving their relative position, not because of arbitrary barriers placed in their way by privilege, discrimination, or nepotism, but because of the objectively inferior performance profile of their bodies and minds.

What we see emerging here, therefore, is a kind of caste system, but not one based on external social status, like the stratified system that long prevailed in India, from Brahmin to untouchable.[3] Instead, these twenty-first-century castes would be based on tangible and wide-ranging disparities of capability. Even if the members of such a society resisted the temptation to openly classify people into corresponding castes, the result would still be a steep de facto hierarchy, ranging from power and privilege at the top, to relative vulnerability and stunted opportunities at the bottom.

With each passing generation, moreover, this gap between haves and have-nots would further widen. If you are one of the lucky few who have access to potent enhancement technologies, then you may choose to augment the capabilities of both yourself and your children; your enhanced offspring will thus possess the heightened capacities that allow them in turn to gain even greater resources and power, affording them the means to further boost the traits of their own children, and so on, in a self-reinforcing cycle through the generations. On the other hand, if you don't have access to the enhancement technologies, tough luck: you and your descendants remain stuck at the bottom indefinitely, with increasingly little hope of breaking out.

I seriously doubt that the citizens of the industrialized democracies would tolerate the emergence of such a biologically based caste system. It runs counter to the egalitarian premise on which democratic societies are founded: that all citizens are equal not just in their basic dignity, but equal also in the right to develop and express the full potential of their personhood. Thomas Jefferson, in the last letter he wrote during his life (quoted in the epigraph to this chapter), vividly expressed his scorn for the structurally unequal societies of the past: saddles on the backs of the many, boots and spurs for a select few.[4] If privileged access to enhancement technologies unavoidably begets the kind of absolute inequality described by Jefferson, then the only solution is either to ban the technologies outright or to find a way to render them available to everyone. Since an across-the-board ban would be unlikely to succeed (for reasons I'll discuss later), this leaves universal access as the only plausible way forward.

Every nation will no doubt go about the institutionalization of universal access in its own way, according to its own cultural and political traditions. Some will rely on a single-payer model financed through taxation, in a manner akin to the health-care and university systems in northwestern Europe today. Others will emphasize the private, free-market model of insurance typified by the American system. Still others may experiment with hybrids of both, offering combinations of subsidies, tax breaks, and low-interest loans. But the basic premise is clear: any society whose citizens reject the caste system described above will need to offer its members equal access to a variety of the most potent and effective enhancement technologies. (I'll discuss the practical challenges of implementing universal access a bit later.)

It is possible that in some nations the ruling groups will resist offering such universal access. I suspect that this would sooner or later result in widespread rioting, upheaval, and even revolutionary violence: it is hard to see how any underclass would long tolerate the kind of biologically based caste system I have been describing. It is one thing to be poor; it is quite another to see yourself, as well as your children and grandchildren and great-grandchildren, relegated to the irreversible status of a physically and mentally disadvantaged underperformer. People will fight for equal access, and there will likely be blood in the streets if they are denied it.

THOSE WHO JUST SAY NO

In all likelihood, some citizens will refuse to adopt major bioenhancements of any kind at all. For certain religious groups, any intervention that partially redesigns human biology will constitute a morally unacceptable form of "playing God." In the eyes of others, it will seem grossly unnatural, a violation of the delicate equilibrium of human ecology established over millennia by evolution. Still others will reject it because it smacks of technological hubris and arrogant perfectionism. Whatever the reason, these people will reject enhancement technologies, both for themselves and, in many cases, for their families as well, opting to remain steadfast with Humanity Version 1.0. I will refer to them collectively as non-mods (i.e., those who reject major forms of modification).

It remains unclear how well non-mods and enhanced humans will be able to coexist side-by-side. At first, during the early decades of enhancement technologies, the difference between the two groups may not pose serious problems. But over time, as the technologies grow in potency, the gap between them seems likely to become more and more significant. For the sake of concreteness, let us imagine a specific case. The year is 2058. I am a twenty-four-year-old non-mod named Cathy, applying for a job as a lab technician at a large pharmaceutical company. I worked hard in college and got good grades. This will be my first real job after graduation, if I can land it. My competitor is a nice young woman named Gillian, whose enhancement package is a fairly run-of-the-mill version widely used by large numbers of middle-class families. Gillian looks about the same age as me, but in fact she is thirty-eight. She has already held technical positions at two other companies before this one and comes with strong recommendations from her former employers. She is able to concentrate intensively on her work for eight hours straight, with only a short break for lunch. I, on the other hand, notice that I become weary after many consecutive hours of hard work and need to shift to less demanding projects toward the end of each day. Gillian's mental acuity is such that she can handle seven experiments concurrently, shuttling down the hall from lab to lab throughout the day, keeping each project's statistics and parameters separate in her mind. I am only able to run two experiments at a time. When I try to do more, the details overwhelm me and I make costly mistakes. She runs the robotic machines in the lab directly through her brain-machine interface, which allows her to control all the most complex devices in the various experiments. I still have to use my handheld computer, which means that some machines remain beyond my abilities, and I have to call in a specialized technician to operate them for me. On the day of our job interviews, I have unfortunately come down with a cold and have a splitting headache. Gillian (who hasn't been sick in years) is feeling fine and projects a sparkling, pleasant personality. Whom will our prospective employer hire?

The multidimensional disparity between these two persons, writ large at the societal level, cannot help but result in a two-tiered socioeconomic system. Unless society enacts laws that set up a mandatory quota program, compelling employers to hire a certain percentage of non-mods, it is hard to see how the most demanding and desirable jobs would not routinely go to those possessing higher performance profiles. All the most menial roles would tend to be left to advanced robots or the unenhanced.

But there is an even more basic level at which the enhanced and the non-mods may find it hard to coexist: their immune systems may gradually become incompatible. As health-enhancement technologies grow more potent over coming decades, it is likely that drug-resistant strains of viruses and bacteria will develop. The bugs will adapt, becoming stronger and more resourceful in their methods for overcoming our ever more sophisticated defenses.[5] For the non-mods, this ongoing phenomenon could eventually come to pose a life-threatening danger. If the evolving ecology of microbes among the population of enhanced humans keeps generating ever more potent strains of superbugs, it is possible that some of these über-microbes could easily overwhelm the relatively weaker defensive barriers of non-mod immune systems. Under such conditions, non-mods would face a choice: either to compromise their moral rejection of enhancements, accepting a certain level of advanced immunological augmentation, or to seal themselves off definitively from the population of the enhanced. If the more hard-core segments of the non-mod population were to opt for the latter, they would have to start wearing the kinds of protective suits that researchers use in biohazard laboratories or else segregate themselves in biologically isolated communities, perhaps enclosed under plastic domes.

Let us assume, for the sake of argument, that a majority of non-mods will yield to necessity and adopt the bare minimum of immunological modifications required to allow them to coexist safely alongside the enhanced population. The other core problem would still remain: their performance levels would tend systematically to relegate them to the lower echelons of the economy and social spheres. One possible solution might be for the non-mods to band together in voluntary self-segregation, establishing enclaves in which a parallel economy and society could emerge. Such a non-mod enclave would presumably generate its own relatively self-contained exchange of goods and services, as well as its own schools, culture, and social activities. In this scenario, non-mods would no longer attempt to compete with the enhanced humans of the mainstream population, but would work and compete and cooperate primarily among their own kind.

My mental image here, as one might imagine, is that of the Amish people. They, too, have opted out of the mainstream economy and culture of their country. They have chosen to reject a broad array of modern technologies. And they have managed to sustain their self-segregation with relative success. Nevertheless, the implications are startling: toward the end of this century, people who look and think and behave like you and me will be comparable, in that context, to the Amish of today. In a social and technological environment that is keyed to ever-rising performance, those who choose to remain biologically unchanged will take on the quality of quaint relics—the stubborn embodiments of a bygone era.

■   ■   ■

Standing back from the various arguments I'm making in this chapter, it becomes apparent that enhancement technologies are not the sorts of innovations to which one can casually say, "Take it or leave it: the choice is yours." Once large numbers of persons start adopting them, they exert significant pressures on the remaining population: either join in or else suffer the consequences. The phenomenon is similar to what happened in Europe around 1800, during the early years of the Industrial Revolution. Once Great Britain launched the modernization of its agriculture and (especially) the mechanization of its textile industry, other nations had no choice but to follow suit. The old woolen and handcrafted garments proudly offered by German or French weavers were no match for the cheap cotton clothing produced under economies of scale in English mills. French and German weavers, like it or not, were forced to adopt their English competitors' methods and strategies.[6]

This "challenge and response" phenomenon is not an uncommon one in the history of economies and technologies: one sees it happening every time a new breakthrough renders one product overwhelmingly superior in quality, efficacy, or price to its market competitors.[7] What is unprecedented about the bioenhancement technologies is that, once they become widely adopted, they would engender this same type of challenge-response mechanism among human persons. Where once we spoke of a certain *product* hopelessly outclassing another, now this same logic would also apply to human individuals. We will encounter this phenomenon again and again in the chapters that follow: the blurring boundary between "person" and "product" is arguably a defining feature of enhancement technologies.

EIGHT

A Fragmenting Species?

Cultural Preferences Inscribed into Biology

From so simple a beginning endless forms most beautiful and most wonderful have been, and are being, evolved.

—Charles Darwin, *On the Origin of Species* (1859)[1]

- **VIGNETTE: *ACROSS THE SEA OF JAPAN***

I hate them.

There, I've said it. No matter how hard I try to work with it, the feeling keeps coming back. I meditate, I practice tai chi, I reread the teachings of Lao Tzu and Gandhi, Confucius and St. Francis.

Yet, still, I cannot but admit my failure. I hate them. It seems like it's getting worse, not better.

The Japanese invaded my country in the last century. Manchukuo. The Rape of Nanjing. Unit 731, the Japanese medical officers who took my people, young and old, infected them with deadly viruses, then live-dissected them to study the progression of the disease.

They treated us like animals then, and it's no different today. In fact, today it's even worse. At least we were still human animals to them, back in the time of the world war. Today we're not even human anymore, in their eyes.

I see them, the tourists from Tokyo or Osaka, looking down at us from their fancy buses as they pass through our streets. The unconcealed disgust on their faces.

And yet it's we Chinese who should be disgusted! They are the ones who've turned themselves into semi-robots! All that we Chinese have done with our bodies has been strictly in harmony with nature's laws. Our epigenetic modifications have made us better organisms, truer to our inner nature.

Not like these Japanese cyborgs, their brains and senses shot through with nano-electrodes and microprocessors. Talking to each other telepathically through their implants, their faces impassive, like the hive-creatures they are.

And they're the ones looking down on us! Calling us monsters, deviants, genetic abominations. It's enough to make your blood boil.

I try and try to follow the precepts of my country's spiritual leaders, who say to us, "Do not hate our Japanese neighbors merely because they have chosen a different path. Their culture has always been more open to robots and informatics. We Chinese have always emphasized the way of biology and natural harmony. Don't forget that underneath these surface differences, we are all still human. We are all Asian brothers and sisters."

But that's not the way it feels.

They are not like us anymore—if ever they were like us.

They never apologized for the outrages of the last century. And now, in the twenty-first century, they look down on us and treat us like inferior beings.

There's only so much humiliation you can ask a person to accept.

Forgive me, my teachers, for saying it, but it's the truth.

I hate them.

RESHAPING OURSELVES THROUGH EPIGENETICS

As I described earlier, two kinds of human genetic engineering may become available over the coming decades. One form, germline reengineering, would require making changes to the DNA of individuals soon after the moment of conception. The other method, epigenetic modification, would target the molecular mechanisms that regulate DNA expression (while leaving the underlying DNA unchanged). In principle, both methods could generate powerful modifications to the body and mind of individuals, but the epigenetic pathway would possess two major advantages. Whereas germline engineering would be a one-shot deal, fixed and irreversible, epigenetic modifications would be flexible, reversible, and upgradable over time. Furthermore, while alterations to the germline would have to be made by parents on behalf of their just-conceived offspring, epigenetic modifications would be available throughout a person's lifetime and will therefore result (in most cases) from choices that individuals will be making for themselves as the years go by.

The idea of creating a designer baby through germline engineering may at first glance seem irresistible to certain ambitious couples who are eager to raise the "best possible" child. Such parents, however, would soon find themselves facing a daunting array of challenges. If mistakes or anomalies were to arise in the design process, they would be partially responsible, alongside their doctors and the genetics industry, for the lifetime of difficulties that their irreversibly malformed offspring would face. A second challenge would be to refrain from laying a heavy burden of expectations on their partially designed children. Already today, some parents go off the deep end in this regard, pressuring their kids to become doctors or great tennis players or some other fetishized professionals. Germline design technologies will compel prospective parents to render their preferences much more explicit than ever before, as they choose to boost some traits for their offspring and rule out or play down others. In such a context, it will be harder for them to avoid subsequent disappointment if their kids fail to display the desired phenotype, or do so in ways that defy parental hopes. Some particularly obtuse parents may even sue the genomics companies that oversaw the design process for their children, demanding their money back.

What is more, couples who embark on germline engineering will need to find ways to negotiate successfully on the selection of traits: we can expect this to exert a new kind of stress on marital relationships, resulting in not a few cases of divorce. After all, this kind of choice is not like selecting a car in a dealership or a resort for a vacation: the stakes involved could not be higher, affecting virtually all aspects of the family's present and future. Many individuals lack the interpersonal skills required for working through entrenched differences of opinion on such weighty decisions.

One strategy for resolving such dilemmas might be for the two parents to compromise, trading off traits desired by one spouse for separate traits desired by the other. Another strategy could be for one parent to have the principal say in shaping the traits of the first child, while the other parent would have the major priority in selecting attributes for the second child. In both these cases, however, unprecedented sources of resentment could be expected to come into play. In the first case, I might find myself bitterly accusing my spouse of having chosen some trait in our kid that resulted in an undesired outcome: "You were the one who insisted on extreme musical ability, and now all he does is hang out with the druggies in his rock band all day and night!" In the second case, one might expect each parent to develop an especially strong bond with the child whom he or she had played a lead role in designing. This in turn could introduce powerful stressors, not only into relations between spouses, but into the overall family dynamics among spouses and kids. Would one kid be considerably more "yours" than "ours" because of her trait profile?

From the child's perspective, moreover, germline modification would raise equally troubling moral issues. Knowing that my parents had tweaked my germline, I would naturally want to learn what design criteria they selected for me. I might realize full well that my genes do not by themselves make me who I am. I might clearly understand that my parents' genetic choices could only constitute, at most, partial influences on my character, not precise determinants of my specific traits. Still, unsettling questions might nonetheless arise: Are my preferences, tastes, and achievements the partial result of the predispositions engineered into my being before I was born? If I become an accomplished musician, for example, and love the whole enterprise of making music, I may still wonder: To what extent am I merely playing out the fact that Mom and Dad chose this for me? Is my great joy in making music *itself* the result of certain dispositional factors inserted into my genes by a splice? Is this achievement really my own?

All these profound difficulties, which are inherent in germline engineering, would still exist to a certain extent for families undertaking epigenetic modifications, but they would present themselves in a far more limited and manageable form. Certain epigenetic modifications aimed at boosting disease resistance or physical stamina could presumably be made by parents on behalf of their babies and toddlers, but such modifications would be reversible later on, if the grown child opted not to retain them. These alterations would also be upgradable and would not face inevitable obsolescence, as germline alterations would. Most important of all, epigenetic modifications targeted toward the core traits of an individual's character—personality, affect, mental acuity—could be left to the judgment of each individual after reaching adulthood. Parents would not have to face the moral burden of imposing their own preferences on their unborn children, but could leave this up to each child to decide as she came of age.

This combination of practical and moral advantages will probably lead most families to opt for the epigenetic method (assuming it proves feasible), and to reject germline engineering. The ability to shape one's own character and mental life over time constitutes one of the defining elements of human autonomy. Germline engineering would violate this autonomy in fundamental and unavoidable ways, whereas epigenetic modifications would leave open the door for individuals to exercise their own judgment and choice.[2] To the extent that I am redesigned at all, I will want to be the one who sets the design parameters. Over the course of my lifetime, I will make a series of decisions about which genetically influenced traits I wish to modify in myself. At each phase in this process, I will survey the options available in the market and select the ones that suit me best. Later, as I grow older and my tastes and preferences gradually shift, I may opt to undo some of my earlier choices and undertake new modifications of an entirely different nature. In this way, epigenetic technologies would open up an entirely new dimension of self-discovery and self-fashioning for the human species.

The question then becomes: What sorts of choices will people make—and what will be the consequences?

TRAIT FADS

If bioenhancement technologies come into being in a free-market consumer economy, then it is likely that they will be subject to the kinds of passing vogues and fashions that typically affect most consumer products. The rising and falling popularity of drugs like Prozac or of cognitive boosters such as Ritalin gives us a sense of the patterns in which future enhancement pharmaceuticals will come to be adopted. In a similar way, bioelectronic implants will probably follow the kinds of buzz and hype that we see today with devices such as iPhones and Google Glass. Epigenetic interventions would be equally subject to the vagaries of consumer preferences and constantly evolving market offerings, and can therefore be expected to appear in the form of passing fads.

A suggestive example in today's society lies in the naming of babies. The popularity of certain names waxes and wanes from one year to another: we see the "Emily decade" for girls, or the "Jacob wave" among boys.[3] In a similar way, we may find that basic bioenhancement profiles will follow analogous patterns: the "musical seventies" or the "blond nineties." This sounds frivolous, but it points toward a more significant conclusion: *in such a world, the impact of culture on the shaping of our biological constitution becomes far more pronounced than it is today.*[4]

Consider, for example, a society in which a majority of citizens harbors the sincere belief that women should be more empathetic and nurturing than men, and that men should be more assertive and competitive than women. Once these people are given the epigenetic tools with which to tweak their own character traits, many of them will select personal qualities for themselves that reflect these gender stereotypes. As a result, these traits will become, over generations, increasingly inscribed into the biology of the people in that social order. Whether or not it was actually true *initially* that women were "inherently" or "naturally" more empathetic and that men were more competitive, this will tend to become more and more the case over time, simply because these will be the cultural values that will steer the design choices made by large numbers of citizens. The stereotype will generate actual predispositions embedded in people's epigenomes.

I see no easy way to avoid this problematic outcome. We cannot escape the fact that we are cultural beings whose worldview bears the imprint of the particular social and historical contexts in which we have grown up. These cultural factors cannot help but come into play whenever we make *any* important decisions, including, of course, our choices about enhancement technologies. The result, in many cases, will be a self-fulfilling prophecy in which cultural norms shape biological propensities in potent ways.

HOMOGENIZATION

Imagine yourself driving with your real estate agent through the suburbs of a large American city of the present day, looking to buy a home. You turn in to a recently developed upper-middle-class neighborhood and visit several houses. After a while you realize that nearly every house possesses the same long list of attributes: high ceilings with crown moldings, hardwood floors, designer kitchen, and so on. Americans like to think of themselves as rugged individualists, but the reality of these kinds of neighborhoods suggests something different. At bottom, a great many of us want very similar things.[5]

One would not want to overplay this observation, of course. Many Americans either cannot afford this type of lifestyle or else *can* afford it but are repulsed by it and seek to carve out a less scripted way of life in edgier parts of town. Nevertheless, the sheer statistical weight of this kind of neighborhood within the overall profile of American dwellings cannot be a mere coincidence: it reveals an underlying convergence of tastes and aspirations that characterizes a significant portion—probably a majority—of the population.[6]

Will this "homogeneity factor" spill over into the domain of enhancement technologies? I can think of no reason why it wouldn't. To be sure, selecting the qualities of your body and mind is a very different matter than selecting the features of your home or car, and most people will no doubt understand this perfectly well. Still, what basis would people have for making their trait-design decisions, other than the same set of fundamental values, goals, and desires that also governs their lifestyle choices? How they choose to live and how they choose traits for themselves would both spring from the same underlying core of values. And for a significant portion of the population, those values will dictate strikingly convergent, homogeneous choices.[7]

It would be a mistake to conclude that the persons who result from this process would all look alike or think alike: that would be to fall into the trap of genetic determinism. Even monozygotic twins often end up having very different personalities and capabilities. It is likely, therefore, that the dynamic, unpredictable nature of genetic/environmental causation will continue to ensure a powerful "scrambling" influence on people's phenotypic traits. Having said this, however, the fact would nonetheless remain: a gradual diminution in the underlying epigenetic and phenotypic diversity of the population would probably occur. Some traits will be more prevalent among these people; other traits will have been systematically pushed aside. The visual image of a "generation of clones" is completely misleading, but it *would* be accurate to say that, as a cohort, some striking patterns and regularities would no doubt characterize these men and women.[8]

IDENTITY PROFILES: WHAT TYPE OF PERSON HAVE I OPTED TO MAKE OF MYSELF?

A common mechanism for signaling our distinctive identity to other people lies in the lifestyle choices we make.[9] Thus, for example, I may drive a Prius, wear Birkenstock sandals, and eat vegetarian food: this will project a certain identity profile to the other people in my social world. It would be quite a different profile from the one suggested by driving a Hummer, wearing cowboy boots, and consuming mass quantities of red meat. Of course, these are mere stereotypes, but they set up tacit expectations in people's minds: if a big yellow Hummer pulls into the parking spot next to me, and the driver gets out sporting Birkenstock sandals and a T-shirt advocating gun control, I will be taken aback. My profile categories are being messed with.

Enhancement technologies will add powerful new options to the array of "signature traits" through which this kind of self-profiling is carried out. How you decide to modify yourself, gradually reshaping your own physical and mental trait constellation, will speak volumes about what kind of person you are. Which enhancement pills do you generally take? What kinds of bioelectronic devices have you installed? How many of the latest epigenetic tweaks have you engineered into yourself or your kids? All these choices will convey a sense of an individual's distinctive "enhancement style," adding a new dimension to the ways in which people construct their identity and signal it to others. Precisely because the enhancement technologies will give people a far greater say in sculpting their own personhood, they will also compel people to articulate their preferences more clearly and openly: "I want to be this type of person, not that type." These technologies will therefore accentuate the tendency of individuals to think of themselves and of other people in terms of explicitly codified labels and groupings.

The companies that market enhancement technologies will no doubt be among the first to pick up on this cultural tendency and to take advantage of it. If people are informally using certain labels as shorthand for describing their own enhancement choices, biotech companies will probably respond with packages of bundled enhancement modifications that are tailored to fit those very categories. Thus, for example, if large numbers of people start casually referring to a certain set of cognitive enhancements as a "smart profile," the enhancement industry would most likely offer product lines that reflect this: Smart Bundle, Insight Array, Intelliform, Genius Pack. Informal enhancement categories circulating in popular culture and product-marketing packages circulating in the consumer economy would therefore reinforce each other over time, further propelling the trend toward a sharpening of identity profiles.

PROFESSIONAL SPECIALIZATION

Some careers in the present day require a considerable degree of training, education, and even physical preparation from those who adopt them. One thinks, for example, of athletes refashioning their bodies over time in preparation for the Olympics, or of the years of daily practice that a concert violinist must undertake before auditioning for a symphony position. Over the coming decades, we can expect these kinds of professional specializations to exert a significant influence on the enhancement modifications people choose for themselves. Thus, the athlete would presumably seek out all the augmentations of prowess and coordination deemed allowable by the officials in charge of her sport; the violinist would use a variety of pharmacological, bioelectronic, or epigenetic interventions to boost her dexterity, concentration, or ability to distinguish subtle nuances of timbre and pitch.

Played out over time among millions of job seekers, this trend can be expected to result in a more intensively competitive employment market, in which ever higher degrees of self-augmentation will be expected from prospective employees. Someone interested in becoming a lawyer, for example, will need to demonstrate not only native analytical ability and an excellent educational record, but a keen memory and a razor-sharp intellect (boosted by chemicals and bioelectronics); the ability to work very long hours under pressure (bestowed by pills and epigenetic interventions); oratorical and rhetorical skills (honed by bioelectronics and virtual reality technologies); and a physical demeanor sculpted to convey credibility, authority, and prestige in the courtroom.

Some lines of work, of course, will not require this kind of aggressive pre-specialization. Working at a fast-food restaurant or painting someone's house will probably not demand as much intensive self-fashioning as the higher-paying jobs at the other end of the prestige spectrum. But that is exactly the point: the more desirable jobs will generate an escalating frenzy of competitive enhancements, precisely because they will be better paid, fewer in number, and harder to land. This phenomenon will probably render it harder for people to switch careers, because they will find themselves competing against other persons who will have already modified their bodies and mental faculties in ways that specifically tailor them for excellence in that line of work.

Some parents may attempt to get the jump on this competitive process, deliberately selecting a particular configuration of genetic predispositions for their offspring, in the hope that these boosted germline or epigenetic traits may give their children an edge in landing the more desirable and high-paying jobs when they come of age. Although these parents may be well aware that they cannot guarantee specific career outcomes—you cannot design your kid to be a doctor—they will be gambling that their children will accept a genetic preorientation toward a particular career path or narrowed set of career options, in exchange for the financial rewards and high prestige that such careers would offer.

Over time, this phenomenon may generate a rather disconcerting side effect: a clustering of people into homogeneous cohorts defined by professional niche. These people might still look and behave quite differently from each other, as individuals, but when you stand back and take them all in, you would be struck by their physical and mental commonalities. To be sure, one already sees signs of this phenomenon today: if you go to a dinner party where most of the guests are doctors, the tone will be quite different from a party where most of the guests are musicians and songwriters. Enhancement technologies will accentuate this tendency, perhaps dramatically so.

A NEW DIMENSION OF PREJUDICE

A bigot, in today's society, has no choice but to base his hateful stereotypes on relatively superficial attributes like skin color or on general behavioral categories such as sexual orientation or religious affiliation.[10] But here the bigot runs into a basic problem: most of the malicious generalizations he wants to make about people are manifestly untrue. Thus, a racist may assert that black people are dumber than white people, but the rest of us can see for ourselves that this is nonsense. The inherent inaccuracy of the stereotype automatically deflates its power, at least in the eyes of people who are willing to observe impartially.[11]

In a world of enhancement technologies, however, all this may change. Some subgroups of people may come to be *truly* different from most others in unprecedented ways. Depending on the profile of modifications they choose, they may have real capabilities that set them apart from most other humans: pharmacological habits that are strange or extreme, powerful bioelectronic implants and prostheses that few others have adopted, epigenetic propensities that endow them with genuinely weird biological features. The divergence from other humans, far from being an "invented" or deliberately exaggerated kind of difference, will be real, visible, and impressive. When you walk down the street and encounter one of these highly modified individuals, casually doing the extraordinary things of which she is capable, the overall quality that will strike you most vividly will be her radical otherness.

A new dimension of prejudice can be expected to emerge at this point. Precisely because it is based on real inequalities in performance, biology, and embodied technologies, it will be more potent and intractable than the prejudices of the past (which were already vicious enough to tear societies apart). People of goodwill who seek to transcend these new kinds of difference will have to work harder to demonstrate the underlying human commonalities that bridge the members of "us" and "them." They will have to show that, despite the undeniable disparities between our trait profiles and theirs, we all still share enough qualities to find a way toward the deeper human identity that embraces us both.

These new prejudices will probably fall into two basic categories: one for the division between non-mods and the enhanced, and one within the world of the enhanced themselves. It is hard to say which will be the more intense of the two. The chasm that separates non-mods from "mods" will be wide and deep, of course, but precisely for that reason, it may actually prove easier for the enhanced majority to regard the non-mods with a measure of benevolence and respect, as many Americans do with the Amish today. As long as the non-mods accept their separate (and subordinate) place on the fringes of society, it is conceivable that "live and let live" would be achievable.[12]

Within the enhanced population, unfortunately, the opportunities for resentment, negative stereotyping, and rising cycles of hatred will be considerable. Let us consider for example a single trait: intelligence. In today's world, if you are a lot smarter than me, I may not like it, but I can make my peace with it. It's nobody's fault, it's just the luck of the draw: I shrug, and get on with my life to the best of my ability. However, when this kind of cognitive disparity is the result of enhancement choices, I will be more tempted to resent the difference between us. The acuity-boosting drugs you take may be stronger; your bioelectronics may be more effective than mine; my epigenetic alterations may be less potent than yours. Whatever the reason, I will be continually confronted with the fact that this was not just a matter of blind luck: I will always know that *it could have been otherwise.* My disappointment at my relatively limited abilities will naturally flow into frustration, humiliation, envy, and anger—especially if you flaunt your superiority.

Under such circumstances, it would be perfectly understandable (though, of course, not commendable) if my psychological reaction were to lash out at you.[13] I might tell our mutual acquaintances that you are an egghead who lacks common sense, that you aren't in touch with your emotions. I might portray you as an abnormally, almost freakishly, one-sided individual. I would also be tempted to characterize all people who have your performance profile with this same negative judgment, thereby generating the seed of a new vilifying stereotype to circulate within the broader culture.

When we survey the sheer number and variety of enhanceable traits that would be liable to generating this kind of corrosive phenomenon, the result is sobering. Disease resistance, mental acuity, youthfulness, ability to interact with machines, capacities for bioelectronic communication with other people, attractive qualities of personality and character—in all these cases, some people will no doubt feel that they have gotten the short end of the stick and will go through a rising spiral of negative psychological reactions. Jealousy and resentment, moreover, can easily spill over into still stronger emotions such as revulsion and outright moral condemnation. I might find myself earnestly concluding that your modified embodiment is weird or grotesque, or that your thoughts and feelings no longer operate within recognizably human parameters. The logical culmination of this escalating series of judgments would be to conclude: You are not one of us anymore. You are a freak, an abomination.

This is a scary prospect. I would be inclined to dismiss it as extreme and unlikely, if it weren't for the fact that human beings have *already* shown a tendency, throughout their history, to engage in precisely this kind of aggressively dehumanizing portrayal and treatment of each other. One thinks, for example, of the civil war in the former Yugoslavia during the 1990s: the stunning rapidity with which longtime neighbors and friends suddenly divided along ethnic and religious lines; their breathtaking willingness to hate and to demonize; the voluptuous fury with which they turned on one another with torture, gang rape, and murder.[14] All this among people who were not really, at bottom, very different from each other.

If the Yugoslavian story were a rare exception, it might not be so troubling in its implications. But the opposite is the case: the long human track record is pockmarked by countless ravaging divisions and conflicts of this sort. What this unfortunately suggests, in the context of enhancement technologies, is that the peacemakers, conciliators, and bridge builders of the coming century will have their work cut out for them.

■ ■ ■

Wait a minute, you may be thinking, as you look back over this chapter: there's a big contradiction here! First I argued that trait fads and convergent tastes will gradually render people more *alike* in various ways. Then I took a seemingly opposite tack, speculating that the identity profiles of individuals and groups might grow sharper and more distinct, and that profound *differences* in people's performance profiles might increase resentment, prejudice, and hate. The contradiction disappears, however, if we take into account the tendency of people to sort themselves into separate categories and factions. What we then have is a scenario of humankind increasingly divided into a wide variety of subgroups and clusters, with greater homogeneity *within* clusters and heightened difference *among* clusters. In this sense, the paradox of enhancement technologies is that they will pose two seemingly contradictory challenges at the same time. Human individuals will have to struggle harder to assert their distinctiveness within their cohort, while groups and collectivities will have to struggle harder to recognize each other's common humanity.

These technologies, in short, will tend to exacerbate the overall fragmentation of humankind. The process will no doubt be gradual, almost imperceptible, as the decades go by, but it will create new kinds of division and tension among human communities. Later on I'll discuss some of the policies that will be needed to keep these centrifugal forces from tearing our society apart.

NINE

If I Ran the Zoo

Adventures along the Plant-Animal-Human-Machine Borderlands

The sleep of reason produces monsters.

—Francisco Goya (1799)[1]

■ **VIGNETTE:** *JEREMY*

Susie had the tact to start with the milder cases. He followed her into the first low building, through a set of double doors and into a long hall with dozens of stalls of different sizes opening along either side, high partition walls dividing them off. Skylights overhead. The smell, first thing, so strong it nearly made him gag. A tangy, musky smell, not of excrement, more like sweat.

She turned into one of the first stalls, where a translucent subceiling shaded the space below. "These are the green pigs."

David stared at them. Six smallish pigs, snuffling about in the hay. They appeared normal, apart from their lime-green color. She explained: "They have chlorophyll engineered into their skin pigment. Only eat about a third as much as regular pigs, because they directly convert sunlight to usable energy. They're basically pork chop plants. Problem with these guys is, their chlorophyll levels are way too high, so they absorb too much energy if they go out into full sun. Just fries 'em. So we have to keep 'em in here under the UV filters."

David observed the creatures. "They don't look too unhappy," he said.

"They're not. Isaiah, there, the smallest one, is getting beat up pretty bad, so we may need to separate him out. But other than that they're doing fine."

"Let me guess: my first important job is going be to shovel out a week's worth of green dung."

She laughed, a pleasant clear laugh. "We've got bots for all that. Our jobs are much harder: making sure all these guys don't suffer too much."

"I thought here in the enclave hardly anybody has bots."

"That's true. We avoid 'em as much as possible. But for things like cleaning pig shit, you can't beat 'em."

He smiled. This was promising. They weren't fanatical here about rejecting technology, like in some of the other non-mod enclaves. They just wanted to keep it in its place.

He followed Susie on to the next stall. Three cats, lying listlessly on a folded blanket. A thin line of spittle hanging from their mouths. "They're heavily sedated," she explained. "Someone decided there'd be lots of money in it if they could fashion cats that were more like dogs. You know, wag their tails, fetch. So they engineered these three females with lots of dog genes. They're pretty much psychotic."

"Why? What do they do?"

"Hump anything that moves, constantly, nonstop."

He laughed. "I know some guys like that, back in Seattle."

Then he saw her face.

"It stops being funny, after watching 'em do it the fiftieth time in a row."

On to the next stall. Mice with nervous twitches. The whole cage swarming with their movement, jerking, quivering, trembling, spasming.

And the next. A featherless crow, its baggy gray skin covered with sores. Eyeing David soberly from its perch.

A snake with six pairs of eyes running down the top of its head. All twelve eyes functional, Susie said.

A pair of she-goats, lying on their sides, their udders huge and bloated like pink inner tubes.

Four whippets, reengineered for racing, their musculature grossly bulging around their limbs, dwarfing their tiny heads. "Really bad arthritis," she said.

"How can anyone justify doing this?!" David burst out.

"Easy. In this case, the researchers said the things they learned by developing these whippets helped cure lots of humans with muscular degenerative disease."

He stared at her.

She shrugged. "It's probably true."

"So who pays for all this?" He gestured around at the building.

"It's a nice little deal. The outside world sends us its animal by-products, the ones that survive. They pay the enclave a hefty fee to take care of them humanely. So their conscience is off the hook. And we get hard currency from the arrangement, which we badly need since our economy is always about to collapse." A wry expression on her face: "We're atoning for the sins of the rest of humanity."

Next stall. Schizophrenic rhesus monkeys, their corpus callosum scarred by failed brain-machine interface implants.

Macaw with part-human proto-pharynx.

Hermaphrodite cow, a postmenopausal cloning experiment gone awry.

She led him toward a side door. "There's one more guy I'd like you to meet before lunch. Over in that small red barn over there. His name's Jeremy."

He followed her into the barn. Same musky smell, only more pungent. A dappled light coming through the skylight. Steel bars, the whole place a cage. A thatched hut over in the far corner of the cage, woven from sheaves of grass and leaves.

"Jeremy," she called out. "You have a visitor."

Silence.

David peered into the darkness, his eyes adjusting. A holo screen on the wall lit up. Letters began appearing, forming words.

GO WAY.

David looked at her, but she ignored him.

"Come on, Jeremy. Just a few minutes so you can meet your new friend David."

He noticed she was speaking more slowly and clearly than usual.

Silence. Then letters.

SLEPING.

"No, you're not. You're just being unfriendly."

THRO POOP.

"You better not! If you want dinner today."

Rustling, the grass parted, and he came out. An orangutan. About one-and-a-half meters tall, a huge round face, round brown eyes. He stood leaning forward, holding a large wireless keyboard in his left hand.

He paid no attention to David, looked only at Susie.

"Can you come forward and say hi?" she asked.

He shuffled through the hay, sat down heavily in front of her, placed the keyboard across his legs. Still no eye contact with David. Like he wasn't there.

"Say hello," she whispered to David.

David whispered, "Hello, Jeremy."

Tap tap tap tap on the keyboard.

CANDY NOW.

She went over to a wall cupboard and pulled out a box, grabbed a candy and came back, holding it up to the bars.

He reached up slowly and took it. His fingers nimbly opening the wrapper, he popped it in his mouth.

"See, if you're nice, Jeremy, then people will come talk to you more. Then maybe you wouldn't be so lonely."

Jeremy took a deep breath, the sound heavy in the room, turned and looked at David. Those eyes, that expression. David felt as though the great creature were looking through him.

"Ask him something," she whispered.

What do I say? he thought to himself, panicking.

The huge face intent on him.

"Are . . . are you happy, Jeremy?" First thing that came into his mind. He knew it was wrong as soon as he'd said it. He glanced at her, but she was watching Jeremy.

Deep slow breaths from the cavernous chest with its tufted red hair and sagging breasts.

Tap tap.

NO.

Silence, the deep breaths.

Tap tap tap tap tap.

AR YOU.

David looked at Susie, then back at Jeremy. This can't be happening, he thought. I'm standing here talking to a giant red ape.

Susie touched his arm. "We have to go now," she said.

He nodded. Suddenly realizing he didn't want to leave.

"Goodbye, Jeremy."

He could see him, as they went out, still sitting motionless in front of the bars.

They walked in silence across the sunbaked courtyard, the sudden brightness dazzling.

"Well?" he said.

"Well what?"

"For Christ's sake. Please explain what I've just seen."

She spoke quickly, her voice flat. "About thirty years ago the United Transhumanists managed to get a bill passed suspending the ban on engineering higher apes. Big rush everywhere of experiments in enhanced ape intelligence."

David nodded. "I've read about that."

"Well, you probably also read that none of the trials succeeded. Then, two years later, the Liberal Christians got the ban reinstated. No more super-apes after that."

David waited, watching her.

Susie frowned. "Jeremy's one of the results of that two-year period. Twenty percent human cognition-related genes, highly boosted mental powers. The psychologists say he's got an IQ that's a human baseline equivalent between 80 and 90."

"So . . . why's he here? Why wasn't he considered a success?"

"Because he's miserable, that's why. He's tried three times to commit suicide."

MIXING IT UP

Most people in the present day remain unaware of the extent to which the age-old boundaries among plants, animals, humans, and machines are already routinely being blurred.[2] Occasionally, one of these newly created beings is so bizarre that it makes headlines: a tomato with antifreeze genes spliced from an arctic flounder to boost its resistance to frost (1991);[3] a rabbit that glows in the dark, thanks to fluorescence genes borrowed from a jellyfish (2000);[4] a robo-rat whose movements can be controlled through a wireless interface with its brain implant (2002);[5] tilapia fish genetically reengineered to produce human insulin (2004);[6] a mouse whose brain contains 1 percent human neurons (2005);[7] "cyborg" rat heart tissue in which every cell is individually connected to a computer by nano-wires (2012).[8]

The biologists' technical terminology rests on the following distinctions: a *hybrid*, such as a mule or beefalo, results from crossing the sex cells of different species. A *chimera* is an organism whose body tissues come from different species; examples include grafted grape plants commonly used by winegrowers, as well as the mouse with 1 percent human neurons in its brain. A third category comprises *transgenic* creatures, whose germline has been modified to include a portion of DNA from a different species; examples include the antifreeze tomato I described above.[9]

This all sounds like "mad scientist" material. In all probability, however, such forms of interspecies mixing will become an increasingly normal feature of tomorrow's society: foreign DNA is coming your way. If biologists were to discover, for example, that a certain species of trout possessed an uncanny resistance to cancer, one would expect medical researchers to work feverishly to identify the sections of the creatures' DNA responsible for the trait and to seek ways to adapt those DNA sequences effectively to humans. Some parents might balk at the idea of inserting "trout genes" into their offspring, but I suspect that many others would easily override such squeamishness, if the modification had a proven safety record and conferred robust resistance to many forms of cancer. Most genetic interventions, moreover, will not derive so straightforwardly from a single species: they will either be wholly synthesized de novo in the laboratory, or they will result from a great deal of complex reshuffling and reengineering of DNA extracted from a variety of species.[10]

ANIMAL SUBJECTS AND THE CONCEPT OF "UPLIFT"

Every major human augmentation—whether through drugs, implants, epigenetic tweaking, or DNA splicing—will first be tried out on laboratory animals. Some of these experiments will result in failure, but others will eventually result in success—a worm, rat, dog, or ape whose traits have been boosted in some novel and potent fashion. The result will be a variety of mighty strange critters, the most advanced of which may approach human levels of sentience, affect, and cognition. Can we justify creating such beings? And once they are among us, how will we treat them?

Take, for example, the transgenic orangutan named Jeremy in the story above, who has been given the Janus-faced gift of augmented intelligence. This has conferred the ability to converse with humans, but it has also given him the ability to become aware of himself, and of his existential situation, at a whole new level.[11] Although this scenario is fictional, it is not at all farfetched: many higher apes of the present day who have been afforded the opportunity to learn sign language and to share the social world of humans have shown somewhat similar mental capabilities.[12]

In science-fiction and transhumanist literature, what was done to Jeremy is known as "uplift." It refers to any situation in which an advanced civilization deliberately sets about elevating the culture or biological traits of a "lower-level" society or species.[13] From one perspective, this might seem like a fundamentally generous undertaking that opens up radically new vistas for the "upliftee."[14] A creature like Jeremy possesses cognitive abilities and an affective range far beyond anything an ape has ever experienced. But is this augmentation of powers necessarily an *improvement*? By what standards would such a judgment be made?

Consider the jellyfish, for example. It is a very primitive creature: no brain to speak of, barely any nervous system. It just floats there, pumping water, ingesting nutrients. But by another set of criteria, it is an awe-inspiring entity: the subtle iridescences of its transparent flesh, the effortless grace of its movements, the miracle of its reproduction from generation to generation. A conclave of the world's most brilliant engineers could not manufacture such a thing.

What would it mean, in practice, to uplift a jellyfish? Would we want to make it faster? Smarter? More energy efficient? We certainly could analyze such a creature from this perspective, assessing it according to all manner of human values and criteria. But that would be to miss the point. The jellyfish is excellent in its own terms, entirely in its own right. Here lies the key weakness of the uplift philosophy: in selecting one species as a target of "elevation" at the hands of another, its proponents may *think* that they are drawing on a universal, species-transcending scale, but in fact no such scale exists. What they are really doing is imposing their own egocentric values onto other creatures.[15] In the end, it sounds a lot like the so-called white man's burden of the nineteenth century: "We Europeans are extremely awesome, and the proof of our awesomeness is that we are even willing to bestow upon you poor savages the generous gift of making you more like ourselves!"[16]

Welcome to the emerging reality of animal enhancement in the twenty-first century. The ability to create advanced creatures like Jeremy is not far away now—for the cognitive augmentation of lesser mammals like mice and monkeys is already well underway. Jeremy is only a fictional invention, of course. We have no way of knowing (yet) what such quasi-human creatures will actually be like, or what it will feel like, to them, to exist as modified beings in this manner. Once we create them, we will have to ask them.

For the sake of argument, let us assume for a moment that the enhancement of great apes such as chimps, orangs, and gorillas will be universally banned (some nations have already enacted such laws).[17] This would still leave plenty of room for animal experimentation in an uplifting mode. Ravens, monkeys, dogs, dolphins, cats, elephants—to elevate the affective and cognitive range of such already impressive creatures is effectively to invite them into a world of humanlike personhood.[18] Already today many dog owners and cat owners swear that their pet possesses genuine companionate attributes: he is loyal, affectionate, protective, playful, remorseful, inquisitive.[19] Even, in some cases, vengeful (I am thinking here of a particular cat I once knew named Ashley).[20] Imagine a situation in which such creatures had their mental traits significantly boosted. Some of them might acquire the ability to communicate their experiences to us through a variety of ingenious devices we will have invented for precisely that purpose. What will be our reaction, if the thoughts and feelings they express suddenly convey a distinct impression of "somebody home"?

By the mid-twenty-first century—if not sooner—this ethical challenge will be upon us.[21] We will need innovative laws to define the rights of augmented animals and their proper place in society (a topic I take up in detail later). How we treat these new arrivals in the realm of personhood (or quasi-personhood) will provide a good measure of how humane we humans really are.[22]

■ **VIGNETTE: *AN ILLEGAL PERSON***

Her friends all called her Pseudo, but her real name was Marie-France Dumesnil. She stood just over two-and-a-half meters tall, towering over all others like a female version of a Viking warrior-god. But it was not her height that rendered her unique. She was a broadband interspecies parahuman, the first ever to survive into adulthood.

Twenty-two years before, in February 2068, at a genomics laboratory in the Universidade Técnica de Lisboa, a scientist named Raul Teixeira had embarked on a series of ambitious animal-human gen mixing experiments. Everything rigorously controlled, everything cleared by the UN oversight authorities and perfectly legal: all embryos were destroyed by lab personnel by the twenty-first day of artificial gestation. Then, in November 2068, one of Teixeira's lab assistants had found a way around the elaborate security systems and stolen one of the embryos, replacing it with a dead embryo he'd set aside from a previous experimental batch. He placed the live embryo in a portable incubator vial and took it home.

From there, over the following days, the embryo changed hands several times, its cash value going up considerably with each black-market transaction. It was implanted on February 27, 2069, into the uterus of a woman in Paris named Adélie Dumesnil. In court documents at the trial that followed two years later, Dumesnil's lawyer tried to lay the basis for an insanity defense. Adélie Dumesnil, he maintained, had been obsessed for her entire adult life with animal-human chimeras, and had made it her life's purpose to be the first woman to give birth to a true parahuman hybrid.

She had succeeded. On November 11, 2069, at Paris's Ambroise Paré Hospital, she gave birth to a healthy baby girl, whom she named Marie-France. She had managed to avoid all the mandatory gestational genetic tests by claiming adherence to the Fundamentalist Congregation of the International Church of Christ, Scientist. The birth was by caesarian section, because the baby was exceptionally large: 6.14 kilos. Dumesnil adamantly insisted that no blood tests be performed on the child, but she was overruled after the doctors examined the baby.

A judge was summoned, as well as a Catholic priest. Marie-France Dumesnil's physical characteristics were catalogued in detail on the hospital's patient database. She had normal human facial features, but the color of her eyes' irises was bright red. Her pale white skin was covered with soft, flexible scale-like plaquettes similar to those of a snake. She had the intact sexual organs of both a male and a female human. Her hands each had eight fingers, the middle three of them slightly webbed together. Tucked away behind each of her ears, the doctors discovered what appeared to be a sensory proboscis, folded into a spiral curl. When she cried, the proboscises would simultaneously unfurl, left and right, to an extended length of 5.3 and 5.4 centimeters.

Blood was drawn and a genetic profile determined. The baby was found to be 88 percent female, possessing more than 6 percent of her protein-coding DNA in nonhuman genetic material. This amounted to a sixtyfold violation of the Human Interspecies Mixing Act of 2037, which allowed no more than one-tenth of 1 percent of a person's genome to incorporate elements drawn from other species. Preliminary sequencing found major genomic segments from at least thirteen species: dolphin, ferret, lemur, snail, cobra, butterfly, blue jay, triggerfish, planaria, eucalyptus, daisy, yeast, and parvovirus H7G13P3R1.

The Catholic priest consulted with the Vatican, then left the hospital. "The creature is not human," he told the clamoring reporters outside. "I cannot give it the sacramental blessing."

In the end, the mother, Adélie Dumesnil, was determined to be compos mentis and found guilty of first-degree premeditated interspecies miscegenation. The crime carried a mandatory sentence of life imprisonment without parole. She appealed, and the appellate court sustained the initial verdict. She was incarcerated at the Prison Spéciale Pour Femmes Cesare Beccaria outside Lyon.

Marie-France Dumesnil's legal status was referred to the Biological Section of the World Court in The Hague. After three months' deliberation, the infant was ruled to be human. The Vatican protested and launched a formal appeal.

She was remanded into the permanent custody of the Aberrant Colony of Saint-Lazare, just outside Marseille. There she remained through the years that followed, her health and psychological development carefully tracked by a large international team of scientists and doctors. Within a year it was evident that the brilliant red pigmentation in her eyes resulted from a defect that rendered her completely blind. She was outfitted with a visual prosthesis, the main sensor array placed in the middle of her forehead as with a normal blind person. The prosthesis gave her full human vision, augmented by zoom, infrared, and ultraviolet technologies.

Doctors also determined early on that her immune system was severely flawed, violently overreacting to certain pathogens, while failing to respond at all to others. They placed her on immune-modulation therapy, administered through a dosing implant above her spleen. But apart from these defects, Marie-France Dumesnil grew up remarkably well in the colony. She possessed a calm, generous disposition and bore the intense scrutiny of all the researchers with good humor. She easily made friends among the hundreds of other genetically aberrant children there and came to be accepted by them as a person like any other. She excelled in her schooling.

And she grew: rapidly and without restraint. By age eight, she was taller than most of the orderlies and medical staff who ran the colony. By age twelve, she had left them all far behind, learning to duck as she went through doorways, reminding herself always to be cautious in playing with the others, lest she break their arms with her huge splaying hands, each the size of a tarte aux pommes, the long fingers radiating like the limbs of a starfish. Other people were fragile and had to be touched tenderly, like eggs or glass.

She had learned at an early age to control her twin sensory proboscises, keeping them tightly furled most of the time beneath her thick blond hair. They were useful, though, in a game of hide-and-seek or during a nighttime walk through the colony grounds: she could hear her friends breathing behind a wall thirty meters away; she could sense irregularities in the heat pattern of the paths in the dark, as she led the way for the others down the allées on a moonless night.

Somewhere around that time, when she was sixteen or seventeen, her friends started calling her Pseudo, in reference to her still temporary legal status as a human/parahuman, which was now going through its fourth appeals case, thanks to the implacable patience of the Catholic Church. She liked the name: it captured well the way she felt about simultaneously fitting in and not fitting in to the human world. Besides, in a place like the colony where everybody jokingly called each other "freak" or "monster," to have a name like Pseudo carried a certain cachet. The name stuck.

FAILURE AND ACCIDENT IN THE MODIFICATION OF HUMANS

It is realistic to assume that, despite all the best efforts of scientists and engineers, some percentage of enhancement modifications will fail to achieve their desired result. The failure percentage may be quite small, especially if rigorous clinical trials precede the marketing of these technologies. Nevertheless, since the overall number of persons seeking enhancements will be large, this unavoidable failure factor will still yield a hapless subset of the worldwide population that is afflicted by all manner of radically new kinds of deformity, disability, and malfunction.

Mental and physical infirmities are nothing new in human society, of course.[23] What will be unprecedented in these enhancement-related afflictions will be the much higher proportion of human design (and design malfunction) that goes into them. Unlike most of the ailments that have affected individuals throughout history, these will not be accidents of nature, but miscarriages of engineering. Some of these failures, perhaps many of them, will be truly awful to behold.

Imagine how many ways things might go wrong with bioenhancement interventions: not enough of the intended trait; far too much of the trait; incorrect trait altogether; deformation or distortion of the trait; improper timing or performance of the trait; unanticipated side effects of the trait's functioning; improper integration of the trait with other key elements of body or mind. For example, one might imagine a man whose capacity for empathy had been accidentally augmented to such a high level as to become dysfunctional: he would be incapable of interacting socially with other people because every aspect of their feelings and moods would be so intensely coexperienced by him as to make him continually break down into spasms of uncontrollable emotion. He would therefore become not only psychologically unstable, but also profoundly isolated. His doctors might give him medications to damp down his affective deformity, but one suspects this would come at a steep cost, leaving him the unpalatable choice of being either trapped in his demented excess of feeling or else semi-stupefied. Failed enhancements of this sort may thus generate not just greater *amounts* of human suffering, but new *kinds* of suffering. The ability to feel, think, sense, and perform at entirely new levels, and in entirely new modalities, may bring with it the concomitant ability to experience misery and anguish in correspondingly innovative ways.

Not all disturbing outcomes, moreover, will result from accidents or design flaws. In the foregoing vignette, I underscored the complicating factor of deliberate human malfeasance: ingenious efforts by fanatics (or just plain irresponsible jerks) to circumvent the restrictions that policymakers will have put in place over enhancement technologies.[24] The creation of strange entities like Marie-France Dumesnil will probably be relatively rare, since it will no doubt carry severe legal penalties. Nevertheless, over the decades to come, it is likely that a variety of individuals or groups who are sufficiently determined in their Frankenstein fixation will find a way to bring into the world their own version of the "perfected"—radically transmogrified—parahuman being. I wonder whether to worry more about the societal effects such an entity will have when its existence is revealed, or to be more concerned for what it will feel like to be that poor creature.

Part III

IDENTITY

· · · · ·

TEN

Mechanization of the Self

There is no delusion more damaging than to get the idea in your head that you understand the functioning of your own brain. Once you acquire such a notion, you run the danger of moving in to take charge, guiding your thoughts, shepherding your mind from place to place, *controlling* it.

—Lewis Thomas (1983)[1]

■ **VIGNETTE:** *PILLS*

Juan looked out across the bay toward Sausalito. Here in Pacific Heights you got a straight view over the water.

The pill he'd swallowed two minutes ago was already taking effect, he could tell. Sunlight always seemed a little brighter on it. He smiled grimly to himself. That's why it's called Lighten.

Problem with Lighten was, you could never really tell if it was you or the chemicals pushing your thoughts along.

Then he frowned. Who gives a damn, as long as I can escape a while from this nightmare. Maria, always Maria.

He strode back to the bathroom, grabbed the pill container, and popped another Lighten. The last one. This would make him edgy, he knew, so he opened the cabinet and took down the bottle of Ease. Popped a half one of those, holding it under his tongue. He knew from experience they made a nice little cocktail.

Almost immediately he felt his jaw relax a bit, the muscles in his shoulders releasing.

Amazing.

He glanced at himself in the mirror. He looked older than he felt. It always surprised him. Who's that middle-aged dude staring back at me? The droopy eyelids over dull brown eyes, the slightly pale strip above his lip where the mustache used to be. Until last week.

He smirked. The face in the mirror smirked back. This made him smile, and the face smiled back.

Whoops, feedback loop.

He went out to the fridge and got himself a beer. The lovely sound of the top coming off. Psscht. Three's company. Lighten, Ease, Blue Moon. And me. A foursome.

Maybe I should call her, he thought.

No. Bad idea.

He went over and sat on the couch, taking in the view across the bay. *Porteous*, he commed.

The machine came rolling in.

"Bring me a Wistful."

"How many milligrams?"

"Twenty."

The machine rolled out, then returned a moment later, proffering the small red pill.

Juan popped it. It tasted like raspberries.

Squinting through the brightness, he could just make out the slender white shape of the Berkeley Campanile over there. It had been rebuilt after the quake of '52. He waved his prosthesis arm over the view, zooming and darkening the image to see the old tower more clearly, then switched back to normal scape.

Suddenly he was thinking about Maria. He remembered sitting with her under that same Campanile, one afternoon way back before the quake. They'd gone over for a movie. Couldn't recall which one. Then, toward evening, they'd bought a couple of those major bockwursts from Top Dog. They'd strolled through the campus, no particular aim, finally sitting down on the huge white marble steps below the tower to munch the last bites of the dogs.

A smile came across his face. He was remembering a little snippet from that distant day, something until this moment utterly forgotten.

Maria had leaned over and kissed his neck. Greasy mouth from the bock. But he didn't care.

He'd turned his head, and slowly their lips had come together. One of those long, delicate, intimate kisses.

He leaned his head back on the couch pillow, basking in the memory.

Porteous wheeled back in, saying there was someone named Bill Costello on the phone about buying the Bodega Bay cottage.

"Tell him to call back later. And don't answer the phone unless it comes from Maria or her sister."

Porteous went back out.

The view stretching across the window, the day, minute by minute, going by over the bay. The Wistful was getting a bit heavy on him. Time to change modes. Juan commed: *Bring me a Flow, will ya?*

How many milligrams?

Four. No, make that six.

He chuckled to himself.

Porteous rolled in with a green triangular pill. Juan popped it.

Pipe me in some old reggae. Start with the playlist called Bob.

A musical note opened in the lower-right corner of his visual field. He selected it, and his head began filling with sound. He raised the volume, cranked up the bass pumps. Directly into the auditory cortex, deep, rich, thrumming, powerful. Louder, louder. The bass, he loved the bass. Perfect, now turn on the Omni. Wham, the notes shot through the tactile and proprioceptive cortices, his whole body becoming the resonance chamber, blasting with the rhythm. "Could you be, could you be, could you be loved?" He felt as though he were melting on the couch.

The songs went through him; he didn't know how many. The sun moved abstractly across the daylight, infinitesimally tracking toward the Golden Gate. Tendrils of morning fog were still visible over there, above the coastal hills to the west.

Suddenly the music stopped.

Silence in the room, like a shock wave.

Porteous wheeling in through the doorway.

He shook free from the couch.

"What is it?"

"It's Maria. She's on the phone."

He looked around the room, getting his bearings.

"Tell her I'll be right there."

He got up, strode heavily to the bathroom, swung open the medicine cabinet door, grabbed the bottle of Erase. Popped a whole one, under the tongue.

Immediately, the bright pastel colors started peeling off his reality, fading down into ordinary hues.

The proportions of the room shrinking back to normal.

He shook his head. Leaned over the sink and splashed cold water on his face.

Looked around at the apartment.

OK, he thought. I'm back. More or less.

He commed Porteous: *Patch her through, audio-only.*

"Hi, Maria." His own voice sounded hollow and strange to him.

■   ■   ■

Juan, my protagonist, projects an extraordinary degree of control over his internal states. By means of a variety of pills, he has become a sort of emotional marionette over whom he himself holds the strings. The essence of this type of intervention lies in its predictability—the precision with which Juan is able to generate highly specific mental states on demand. This quality, of course, is the hallmark of the world of machines: it's why we like them so much and depend on them so pervasively. Step on the gas, the car goes forward; step on the brake, the car stops (most of the time, at least). Now I want to explore the implications of bringing this kind of fine-grained control into the deeper recesses of our selfhood.

AUTHENTICITY IN A PILL?

In the early 1990s the psychiatrist Peter Kramer began treating a woman named Sonia, who had come to him with symptoms of minor depression. Sonia was a successful graphic artist, and she struck Kramer as "an ethereal young woman" with a "vague, hesitant habit of speech sometimes characteristic of artists."[2] Her emotional symptoms were relatively mild, but because other members of her family had experienced severe depression, Kramer decided to prescribe Prozac.

Sonia turned out to be what psychiatrists call "a good responder." Not long after she began taking the medication, her depression disappeared, and she also reported feeling more energetic and focused than she had ever felt before. Kramer was particularly impressed by the change in Sonia's speech patterns: her rather hesitant manner of articulating her thoughts gave way to a forceful, cogent way of speaking. She herself noticed the change and very much enjoyed it: she told Kramer she had never before experienced such mental clarity.

After a few months, Kramer felt it was time to wean his patient off the meds. He discontinued the drug, and she reported no adverse effects. Her depression did not come back, and once it had become clear that her return to normal functioning was stable, both she and Kramer concluded that she was cured.

Several more months passed. Then, one day, Sonia came to see Kramer once again. Her mood was fine, she reported: she was not at all depressed. But she was troubled by the gradual dwindling of the mental sharpness she had experienced while on Prozac. She wondered whether Kramer would be willing to resume the prescription for her.

Kramer hesitated several days before making his decision. In the end, Sonia persuaded him that her career and personal life were sufficiently demanding, and her lack of clarity sufficiently troublesome, to justify going back on Prozac. He prescribed a very low dose. Within a short while, Sonia's higher level of function returned. Sharply focused, mentally nimble, she was once again speaking in crisp sentences at a brisk clip. She left Kramer's care, delighted with her "new and improved" condition. Kramer, for his part, was left wondering if he had done a good thing, for he acknowledged that he had crossed a line with Sonia: he had deliberately enhanced the functioning of a perfectly normal mind.

Kramer wrestles with this issue at considerable length in his pathbreaking 1993 book, *Listening to Prozac*. Suppose, he asks, that we could develop "mood brightening" drugs that worked reliably and effectively, were nonaddictive, and had no negative side effects. What objections might be raised against the widespread use of such drugs by healthy persons? The potential problems, he concluded, boiled down to three main issues.[3]

- *Disconnection from reality.* Over eons of evolution, we have developed emotional responses to life's changing circumstances: we recoil fearfully from dangerous situations, relax amiably in the presence of trusted loved ones, grieve over the loss of friends, and so on. Any drug that allows us to override at will the more negative or unpleasant of these emotions is potentially harmful to us because it prevents us from responding appropriately to real dangers or threats. It also shields us artificially from aspects of ourselves that we need to know about, for our experience of frustration or inner unhappiness is often a sign that we need to alter our behaviors or thought patterns in some way.
- *Impoverishment of our experience.* A mood-brightening drug, if relied on regularly, would numb us to some of the more valuably poignant and vibrant aspects of life. We might be more contented people on such a drug, but our experience would be shallower, the range of our sensations and perceptions much narrower, if we could selectively edit out feelings associated with tragedy, horror, anxiety, resentment, ennui, despair.
- *Reinforcement of oppressive cultural expectations.* There is a deep tendency in our society to mask or suppress negative emotions, hiding them away so that they do not trouble the state of perpetual contentment to which the collectivity aspires. Thus, for example, the funeral industry goes to great lengths to insulate us from the harsher realities of death, and the experience of grieving is often stifled or cut short in the name of a return to normalcy. In such a social context, "cosmetic psychopharmacology" would allow people to conform even more closely than today to social norms that proscribe "bad" feelings.

These are all sensible and persuasive arguments, and Kramer readily acknowledges as much in his book. But then he comes back to the case of Sonia, and he asks: Has this young woman's life, on Prozac, really been diminished in any of these ways? Has she become disconnected from reality, numbed to the depth and breadth of life's emotional pageantry, trimmed down with an affective cookie cutter into some culturally acceptable shape? Not at all. She is still in most ways the same person as before, except that she is mentally sharper, more alert. She still experiences the full range of life's ups and downs, and has by no means been prevented from encountering the peak or the abyss. Most importantly, the drug does not induce a state of mind and emotion largely unrelated to Sonia's life events. Rather, the chemical subtly heightens her ability to experience her own customary train of thought, her own ideas, in a more focused and sensitive way. In this sense, it sharpens her familiar sense of herself, rather than distracts from it.

In all these ways, then, we see here a pharmacological intervention that defies the imagery of dehumanization.[4] We see instead a healthy young woman, taking steps to become even more empowered, more richly engaged, more broadly challenged by life's possibilities than ever. Through a pill.[5]

DEGREES OF CONTROL AND PREDICTABILITY

The world of humans is inherently messy. We can be unruly, spontaneous, irrational, erratic. We are constantly surprising each other—and ourselves—with what we say and do and feel. This is why the quest to gain a measure of control in our lives is so fundamental to being human. Much of our energy, day to day, is spent in endeavors that aim to project order onto the people and events that surround us. We create schedules, form organizations, sign contracts, exchange vows—all in the effort to impose an element of predictability on the capricious and changeable activities of our fellow humans. This principle applies to our inner lives as well: our emotions and moods can sometimes feel as though they are running away with us, and we devote considerable energy to developing tools for handling them, whether spiritual, psychological, or pharmacological in character.[6]

But this quest for control can also become problematic if it overreaches itself. Consider, for example, the remarkable discovery made by the MIT researcher Sherry Turkle when she conducted interviews of robot owners in Japan. Some of these people, she found, actually preferred their artificial companions to interactions with living beings! One elderly woman had this to say about her experience with AIBO, the sophisticated doglike robot made by Sony: "AIBO is better than a real dog. It won't do dangerous things, and it won't betray you."[7]

A robot dog—better than a real, living canine? What this satisfied robot owner evidently failed to grasp was what she was losing in the exchange. The possibility of betrayal betokens precisely the kinds of qualities that render a real dog interesting.[8] Your pet poodle does not betray you, partly because it is devoted to you and cares about you, and partly because it relies on you for food and affection. That is what companionship is all about: mutual bonds and mutual vulnerabilities. But it *could* betray you someday. Consumed by a sudden doggish passion, it could run off into the woods with another poodle and disappear from your life. The fact that it doesn't do so, day to day, renders your appreciation of its loyalty even keener. By contrast, the reason why the AIBO robot dog does not betray its owner is because such bonds and vulnerabilities are simply irrelevant to its functioning. Her very certainty that the AIBO can never betray her is precisely what robs the interaction of its deeper sense. Only a machine can offer that kind of uncomplicated, automatic, and utterly predictable performance.

The enhancement pharmaceuticals of tomorrow—far more potent and precise in their effects than those of today—promise to dramatically augment our ability to project fine-grained control over our own mental and emotional lives. For this very reason, however, they may turn out to be "too much of a good thing." Control and predictability are the hallmark of the world of machines, and the more pervasively we inject these qualities into the fabric of our everyday experiences, the greater the risk of crossing a line into the world of AIBO. It is a danger that the philosopher Neil Levy calls "mechanization of the self."[9]

Here lies the key dissimilarity between the cases of Sonia and Juan. (Let us put aside, for the moment, the fact that Sonia's story is based on an actual clinical experience, whereas Juan's is illustrative fiction.) Although both these persons used chemical means to modify their emotional and mental lives, Sonia's use of Prozac did not plunge her automatically into a fixed and one-dimensional mood. Instead, it gently modified large swaths of her personality and cognition, allowing her to engage more fully with the people and events of her daily life.[10] Juan, on the other hand, crossed a line into a world of machinelike causes and effects. When he took Lighten, Ease, Wistful, Flow, and Erase, the results in each case were precise, immediate, and foreseen. This kind of "guaranteed" outcome is excellent in a household appliance or robot. Push the button; it responds the same way every time. In a human being, however, it betokens a dismal fact: he is treating himself like a kind of automat for emotional states. Insert coin, tasty experience delivered.

One can see how it would sometimes prove tempting to simply bypass all the hard work and messy complexity of the human world, and just take the easy route of direct physiochemical control. The cost of doing this, however, turns out to be grievous indeed, for what you are bypassing is your personhood itself. You are temporarily suspending all those wondrous, baffling, frustrating qualities that render you human, and turning yourself into a sort of "experience machine." The realm of compassion and vulnerability, loyalty and risk, sharing and loneliness, becomes less relevant: you have traded it for the straightforward causal realm of billiard balls colliding on a table. The more you do this, the less meaning you will find in your existence, for what you are actually bypassing is precisely what renders a life worth living.

BIOELECTRONIC INTERVENTIONS IN THE BRAIN/MIND PROCESS

In chapter 3, I argued that, by the latter decades of the twenty-first century, humans will probably be able to use bioelectronic devices to modulate their own mental and emotional states with considerable potency and precision. I am envisioning here something akin to what Juan was doing to himself in my story, but operating through the medium of electrical or magnetic pulses rather than chemicals. The device might take the form of a noninvasive brain-machine interface, something similar to a skullcap but much smaller, like a subcutaneous patch. Unlike today's drugs, such a machine would be able to act directly on the brain centers associated with particular moods or abilities. It might therefore cause fewer side effects, such as the jitteriness associated with caffeine or the next-day stupor that accompanies many chemical tranquilizers and sleep aids. It would offer finely grained increments of effect, infinitely scalable and adjustable. It would also be quickly and easily reversible: just push a button and the effect ceases. Back to unmodified me.

Might we become addicted to such devices, coming to depend on them for delivering their satisfying outcomes? Here, one of Robert Heath's experiments of the 1970s bears an eerie relevance (for more on Heath's work, see chapter 3). Heath placed electrodes into the septal region of a female psychiatric patient's brain—a region associated with reward stimuli—and then filmed his interaction with her while a technician delivered the electric pulses. The filmed scene is described in a 1986 book by Judith Hooper and Dick Teresi, *The Three-Pound Universe.*

A woman of indeterminate age lies on a narrow cot, a giant bandage covering her skull. At the start of the film she seems locked inside some private vortex of despair. Her face is as blank as her white hospital gown and her voice is a remote, tired monotone.

"Sixty pulses," says a disembodied voice. It belongs to the technician in the next room, who is sending a current to the electrodes inside the woman's head. The patient, inside her soundproof cubicle, does not hear him.

Suddenly, she smiles. "Why are you smiling?" asks Dr. Heath, sitting by her bedside.

"I don't know . . . Are you doing something to me? [Giggles] I don't usually sit around and laugh at nothing. I must be laughing at something."

"One hundred forty," says the off-screen technician.

The patient giggles again, transformed from a stone-faced zombie into a little girl with a secret joke. "What in the hell are you doing?" she asks. "You must be hitting some goody place."[11]

If a day comes when these kinds of bioelectronic devices become widely available in noninvasive form, they will offer every one of us unlimited access to our own inner "goody places." How will we respond to such a technology? Will some persons choose to retreat into their homes to zap their pleasure centers as unrelentingly and mindlessly as the rats studied in the 1950s by James Olds? Will the technology have to be banned, driven into an illegal underground market like the crack cocaine and meth of today?

■   ■   ■

In the 1936 film *Modern Times*, Charlie Chaplin splendidly captures the essence of what it means to mechanize your selfhood.[12] He works on the factory assembly line, continually repeating the same motions as he tightens two similarly placed bolts on an endlessly flowing sequence of identical widgets rolling swiftly past him on a conveyor belt. The point is not intended to be subtle. Not only does the poor fellow go bonkers, but at the scene's culmination, he is swallowed whole by the machinery, his body hauled through a vast labyrinth of giant interlocking cogs and wheels.

Mechanization of the self, in short, is not a new phenomenon. Whenever human beings have intensively subordinated their individual autonomy to the dictates of a larger system, routine, or organization, it is likely that some of the key dehumanizing features I've described were taking place. One sees signs of it, for example, in military organizations seeking to break down new recruits into effective soldiers, in the drudgery of menial jobs requiring repetitive labor, in large bureaucratic organizations where uniformity of conduct is rewarded over initiative. What all these phenomena have in common is the effort to impose a high degree of discipline on the fickle, erratic (imaginative, spontaneous) creature that is a human person.[13]

Bioenhancement technologies, however, differ from all these examples in one important respect: the locus of control. Charlie Chaplin, in the movie, is eventually rescued from the machine's innards. The Big Machine is external to him: it envelopes him, but ultimately he is able to make his way out and find himself again. He totters off down the sidewalk and the main story begins. In a similar way, if you don't like your job because of the drudgery, you can quit and seek other forms of livelihood. If you grow tired of working in a faceless bureaucracy, you can strike out and start your own business. Because the mechanizing factors lie outside you, in the surrounding environment, escape remains a possibility.

With enhancement technologies, however, the locus of control lies within each person. Once the chemicals are in your bloodstream, you cannot prevent them from having their intended effect: they will cause sequences of reactions in your brain that will induce you to behave or feel in specific ways whether you "want" to or not. When the bioelectronic device transmits its commands to the brain's tissue, there is nothing you can do from within yourself to block them. You will do, or experience, whatever the electromagnetic signals compel.

In short, the ante has been upped. Our society now possesses, or is actively pursuing, the means to extend control directly over the innermost processes that make us who we are, moment by moment. These new powers will potentially confer tremendous benefits. They also carry the risk of erasing our humanity.

ELEVEN

Turbocharging Moral Character

Lionel Merble was a machine. Tralfamadorians, of course, say that every creature and plant in the Universe is a machine. It amuses them that so many Earthlings are offended by the idea of being machines.

—Kurt Vonnegut, *Slaughterhouse Five* (1969)[1]

■ **VIGNETTE:** *THREE PATHWAYS OF ACTION*

Scenario 1 (unenhanced egoism)

You are walking down the street and see an old lady being robbed by a masked assailant with a gun. You think to yourself, "She really needs help, but that guy will probably shoot me if I intervene." You run in the opposite direction.

Scenario 2 (unenhanced altruism)

You are walking down the street and see an old lady being robbed by a masked assailant with a gun. You think to yourself, "She really needs help, but that guy will probably shoot me if I intervene." You hesitate a few seconds, trying to figure out what to do. Finally you grit your teeth, override your misgivings, and run over there, shouting at the man to leave her alone. He sees you coming and runs away. The old lady thanks you profusely. You feel good.

Scenario 3 (chemically enhanced altruism)

You are walking down the street and see an old lady being robbed by a masked assailant with a gun. You think to yourself, "She really needs help, but that guy will probably shoot me if I intervene." At this point, the sophisticated "moral enhancement" pill that you take every morning kicks into action. Oxytocin is released in your brain, and you feel a rush of empathy for the old lady. You run over there, shouting at the man to leave her alone. He sees you coming and runs away. The old lady thanks you profusely. You feel good.

THE MORALITY PILL

In a 2012 *New York Times* op-ed piece, the philosophers Peter Singer and Agata Sagan pose a provocative question: What if we could make people behave much more kindly and thoughtfully toward each other by means of a pill?[2] They observe that neuroscientists are learning more every year about the neural and biochemical mechanisms that underpin our moral judgments. As a result, suggest Singer and Sagan, we are perhaps not far from being able to chemically manipulate those very processes, augmenting the underlying propensity of individuals to make ethically good choices rather than bad ones. Give people a "morality pill," they argue, and you could go a long way toward solving some of society's most intractable problems.

The idea may sound fanciful—an exemplar of the far-fetched hypothetical questions undergraduates sometimes encounter in Ethics 101. But it's not. Neuroscientists and cognitive psychologists are indeed closing in on the mechanisms that underlie empathy, social connectedness, and judgments of right and wrong. They know what parts of the brain are involved, and they are gaining ever greater insight into the neuronal processes that support these mental phenomena.

The neuroeconomist Paul Zak, for example, has conducted extensive experiments on oxytocin, a chemical that has been variously dubbed the "trust molecule" or the "love hormone."[3] Oxytocin serves as a neuromodulator in a variety of important brain processes, ranging from sexual behavior to social bonding. Zak has demonstrated that this chemical rises and falls in direct proportion to people's feelings of connectedness and trust for each other. He has also shown that you can spritz some of this chemical into your nose and significantly increase your chances of feeling warm and fuzzy toward the persons around you.[4] These discoveries, in turn, take Zak into the same intellectual territory as Singer and Sagan: "Is the key to a better society getting everyone to simply snort oxytocin every few minutes? Aside from a number of practical obstacles, the infusion route isn't really necessary, especially since . . . nature provides so many techniques for oxytocin release just in the course of everyday living."[5]

It turns out that every time we give each other hugs (at least, sincere ones), we increase each other's oxytocin levels. The hormone goes hand in hand with just about every social situation in which individuals come together emotionally in fellowship. It can act as both a cause and an effect of making nice. If you and I show trust in each other, this will raise our oxytocin levels; conversely, if a scientist injects me with oxytocin, I will be more inclined to feel trust toward my peers.[6]

We thus return to the proposal of Singer and Sagan. What if we could find a way to infuse people's brains with oxytocin or one of its kindred chemicals, say, by mixing it into the water supply as we do with fluoride for our teeth? Let us put aside, for the sake of argument, the obvious practical objection that, if we make everyone more caring and trusting in this way, then we are setting everyone up to be duped and taken advantage of by those few individuals who will have managed to drink only Perrier and therefore not be oxytocin-infused. Instead, I want to consider the implications *for morality itself* of using chemical means to augment the decency of a person's behavior.[7]

In their thought-provoking book, *Unfit for the Future: The Need for Moral Enhancement*, Ingmar Persson and Julian Savulescu argue that moral enhancement need not necessarily take the form of a pill that made you feel *irresistibly* compelled to behave in an altruistic manner. The chemical they envision would not totally override your free will, transforming you into an automaton of kindness. Rather, it would gently alter your underlying inclinations and impulses, shifting them toward the altruistic end of the normal human spectrum.[8] Such a modification, in other words, would predispose you to be more like the virtuous individuals in our society who—through their inner tendencies of empathy, kindness, or honesty—choose "naturally" to tell the truth or to rush to the aid of those in need.

This is a subtle and intriguing argument, but it fails to acknowledge that the key to the moral meaning of all persons' behavior—whether they are generous or mean-spirited—lies not merely in what they do, but also in *the intentions and choices* that lie behind what they do.[9] Consider, for example, the three scenarios laid out in my story of the old woman being robbed at gunpoint. In Scenario 3—chemically enhanced altruism—how much of a choice do you really have? The extra oxytocin in your brain precludes you from seriously contemplating such selfish actions as running away—for you find yourself awash in a surge of empathetic feeling—and your final decision therefore rests on an exercise of free will that has been narrowed, prechanneled, in a significant way. You really had no alternative but to do some form of "the right thing," because your emotional responses have been chemically nudged toward the kinder end of the dispositional spectrum. The moral significance of your action is therefore greatly lessened, because the *opportunity to fail* was never real and present for you.[10] Behaving nastily or selfishly was never a genuine possibility that you had to decide to accept or reject.

By contrast, the decision in Scenario 2 (unenhanced altruism) is superior in two respects. It entails genuine free will, in the sense of making a reasoned selection among the full range of actions available to you in that situation—nasty, neutral, or nice. And it is an expression of your mental autonomy, because the final choice arises out of your own normal emotional and cognitive processes, without external chemical interference. Your decision to help the old lady possesses rich moral meaning because of these two factors.

This is where the arguments for "moral enhancement" presented by Persson and Savulescu (and Singer and Sagan) fall short. In the name of generating a badly needed improvement in the overall moral profile of our civilization, they advocate a form of bioengineering that would partially restrict the free will and autonomy of individual citizens. But the cure they propose would be worse than the ailment itself. To the extent that our society succeeded in engineering nastiness and selfishness out of the human dispositional repertoire, rendering such unpleasant behaviors less and less a real psychological option for people, we would thereby be moving toward a world without any moral meaning at all. It might be a very harmonious society indeed, but it would be more akin to that of the ants, which all behave in a wonderfully self-sacrificing manner toward the collectivity, but whose behavior lies outside the realm of morality because it does not arise out of free will.[11]

We thus find ourselves, once again, in the AIBO paradox. Precisely insofar as the chemical has rendered our behavior more predictably and consistently altruistic, it has also robbed it of its moral meaning. Even though, when seen from the outside, our resultant actions might *appear* wonderfully generous, they would in fact be originating in mental processes that have little to do with the autonomy and choice that characterize a fully functional personhood.[12]

Philosophers like Singer and Sagan are quite right, therefore, when they observe that "moral enhancement" is likely to become scientifically feasible in the years to come. But our society should resist the temptation to go down that road. The ability to pre-orient people's behavior through direct chemical or neurological means can only come about by reducing their free will. And no societal or personal benefit is worth paying that price.

TWELVE

Shared Intimacies

Monitoring and Transmitting Mental Information

Imagine if a teacher could convey a mathematical proof directly to your brain, nonverbally. . . . We realized that we had all the equipment we needed to build a rudimentary version of this technology.

—Rajesh Rao and Andrea Stocco,
brain researchers (2014)[1]

■ **VIGNETTE: *AURA***

Francesca stretched forward across the coffee table, skewered an olive with a toothpick, and leaned back, her jet-black hair unfurling across her bare shoulders as she looked over at him.

"Have you installed the Aura yet?"

Jed nodded. "Late last night."

"Well?"

"To be honest, I haven't noticed much difference so far. But I haven't been using my comm system with anyone except the bot."

"Let's try it now."

"OK," he said. Slight pause. "You go first."

She glanced at him. He looked so much younger since the last round of juve treatments. His lean face with the perennially earnest expression, the short-cropped hair.

"So . . . what should I start with?"

"Why don't you send me a physical impression. Something simple, like a food taste?"

"OK. You ready?"

He made sure his comm channel was open. "Yep."

A brief pause, then an overwhelming flavor of fresh pastry filled his awareness. It felt like it was emanating from his tongue and mouth.

He grinned. "Cool!"

Wait, she commed, *I haven't finished yet.*

Suddenly he was in Rome, early morning, standing in a street outside a café. At least, it looked like Rome. He was about to sit down at a small red table along the sidewalk and had just taken a bite out of one of the two cornetti in the white paper bag he held in his hand. His fingers had powdered sugar on them. On the tabletop, his cappuccino was steaming, a whiff of the strong coffee and hot milk detectable even without sitting down. He noticed two cigarette butts lying in the gutter. He was in a hurry. A motorbike was roaring past in the street.

"Whoa!" He looked across at her on the couch.

Francesca grinned. "Amazing, isn't it?"

"It's almost as if I were there!"

"Well, you sort of were. That was taken directly from my breakfast last Tuesday."

He shook his head in disbelief. He'd certainly seen all the ads on TV, and he'd assumed it was just the usual hype. But this was remarkable.

"We're just scratching the surface," she said. "The thing about Aura is that all the memories are interconnected. You can find yourself going down a chain of associations, one after another. You know, like Proust. The name "Aura" was lifted directly from neuroscience."

"So . . . I could have traveled down a chain of your memories, if I'd kept going with it?"

"No, I would have had to do the traveling myself and keep transmitting it to you as I went. The Apple people were very careful about leaving control over the memories in the hands of the person they belong to. For obvious reasons."

Hell yes, for obvious reasons, Jed thought. This thing was dynamite.

"OK, your turn now," she said.

Jesus. What if I accidentally send her something from my college dorm days?

"How do I pick a memory to send?"

"Didn't you read the instructions or go through the training session?"

"I never read instructions. You know that."

She rolled her eyes, then walked him through the process. Open Aura. Call up a memory in your mind. Tag it. Aura briefly replays elements of the memory so you're sure it's the right one. Aura asks, "Are You Sure?" You select "Yes." Then it's loaded in the "Send" bay. You hit Send. Done.

He took a deep breath. What on earth do I choose? A lot of options here.

For some reason, an image came up from the day Catherine was born. He spent a moment visualizing it in his mind, then tagged the memory, went through the authorization process, and sent it.

The Aura icon opened in her lower visual field. She closed her eyes and selected it.

"Here she comes," says Dr. Mateus.

Catherine comes out head first, and my first thought is that she doesn't look all that bad. Her skin is a lovely light pink, and she's covered with viscous fluid and a white pasty powdery stuff, but I don't have time to think about what she looks like, because immediately there's a moment of supreme tension in the room as the doctor and the nurse and I wait and watch her flushed little face. Her expression takes on a look of acute displeasure, and her toothless mouth opens, and her hands clench and unclench in the air, and she utters a long gurgling-gagging sound, a sound of frail but profound effort, as she takes in her first breath of air. Then the air comes back out, a miserable little wail, as thin as watered-down tea. Dr. Mateus hands me the scissors and lets me cut the umbilical cord. I find it to be amazingly tough and hard to cut.

I hold her in my hands, out in front of me, this fragile parcel. She's still crying, so I say to her, in a soft voice, "Catherine, Catherine, Catherine." She immediately stops crying, and I find myself looking into her eyes. They are wide open, bright blue, and they are looking at me. Looking right into me.

Francesca opened her eyes. The clock on the mantel said 4:40. Less than an hour had gone by. It felt like it could have been an entire day and night. Or longer.

Jed sat across from her, his eyes still closed. A strange, peaceful expression on his face.

She felt a shiver pass through her. Amazing, amazing.

It had been a torrent of images, feelings, her whole self pouring into it. And the flow coming back the other way, a counter-torrent of this other self, pouring itself into her. This other gentle, funny, self.

She felt totally spent. Hollowed out and filled again.

In that timeless space, they'd sent each other weird dreams, sensuous dreams, the kind you'd never feel comfortable sharing with anyone. Funny memories, the two of them laughing their way through. Miserable, wretched memories, making her want to break in two from the sorrow.

They'd sent each other fleeting little moments, glimpses into passing impressions that have no name. Whose were they? His or hers? Many of these memories she hadn't even realized she possessed anymore. She recognized them as they came up in the cascading connections, one after another, each summoning up another and another, presenting themselves like old friends not seen for too many years.

She closed her eyes again, letting her breath rise and fall, letting the reverberations play across her awareness.

PEEPING AND SHARING

The telepathic communication experienced by Jed and Francesca is intriguing to contemplate, but could such a technology ever become a reality? We will no doubt continue to remain in the dark about this question until scientists make far greater headway in understanding the functional architecture of the human brain.[2] Nevertheless, in a series of studies conducted in 2013 and 2014, three groups of scientists have successfully carried out proof-of-concept experiments for precisely this sort of direct brain-to-brain communication—first in rats, then with human subjects.

The initial breakthrough came in 2013 at Duke University, in the lab of the neuroscientist Miguel Nicolelis.[3] For this experiment, Nicolelis trained rats to press one of two levers in their cage when a light came on. Then he divided the rats into two groups in separate cages: sender rats (or encoders) and receiver rats (or decoders). He placed electrodes into the motor cortices of both groups of rats, linking them through a computer. The experiment consisted of two phases: first, recording the neural activity of an encoder rat's motor cortex while it was performing the lever-pressing task; then, transmitting the recorded pattern into the motor cortex of a decoder rat in the other cage. In the first phase, the implanted electrodes were *retrieving* information from one rat's brain; in the second phase, the electrodes were *stimulating* the other rat's brain with the recorded pattern.

To the wonderment of Nicolelis and his colleagues, the linkage worked. The receiving rat (decoder) was able to successfully replicate the task and press the correct lever between 60 and 72 percent of the time. Neither rat could see the other: all that the decoder rat had to go on was the incoming pattern of recorded neural activity in its motor cortex. Yet this proved sufficient for the rat to know which lever to press.

An equally impressive result was achieved later that year in an experiment at Harvard Medical School: six human subjects in the lab of brain researcher Seung-Schik Yoo were equipped with electroencephalogram

(EEG) headsets and linked via computer to anesthetized rats in another room. Using their thoughts alone, the human "senders" were able to transmit a stimulus that reliably caused the recipient rats to move their tails.[4]

The next proof of concept took place in 2014 at the University of Washington—this time between two humans. Rajesh Rao, a professor of computer engineering, and Andrea Stocco, a professor of psychology, built an apparatus that allowed them to work together, brain to brain, on completing a simple task.[5] In one room, Rao donned an EEG skullcap and positioned himself in front of a video game. In another room, Stocco was fitted out with a transcranial magnetic stimulation (TMS) device whose beam was focused on a particular area of his brain's motor cortex. The two devices were linked over the Internet via computer. All Rao needed to do was to form the intention to push a button. The skullcap picked up the corresponding neural activity and transmitted it to the TMS device hovering over Stocco's head in the other room. The TMS zapped Stocco's brain with a mild magnetic pulse, and Stocco's hand depressed a button on his keyboard. Rao and Stocco reported:

> As we played the game, we got better and better, to the point where in our last run, we intercepted the pirate rockets with almost 100 percent accuracy. Rao learned how to imagine moving his hand in a consistent manner, giving the computer a chance to make sense of his EEG brain data. Stocco found that he did not know his wrist was moving until he felt or saw his hand in motion. We have now replicated our findings with several other pairs of humans. . . . Our next set of experiments will explore targeting other brain regions to produce a conscious thought. For example, we believe we can send visual, as opposed to motor, information from one brain to another.[6]

To be sure, these feats remain far removed from what a sci-fi novel depicts when it portrays telepathic communication between humans. Nevertheless, a threshold has been crossed. We now know that both ends of the transmission process can be made to work: recording synaptic activity in one brain, then producing a corresponding stimulus directly in another brain, with a sufficient level of accuracy to generate meaningful results. (Note that this achievement—while remarkable in itself—is not necessarily the same thing as *faithfully replicating* the original stimulus in the receiving brain, which constitutes a far more arduous and elusive goal.)

Over the coming years, it is likely that the pieces of this complex puzzle will continue gradually to come together. Further research on animal and human subjects will refine the retrieval and transmission technologies. Clinical work with neurological patients will lead to safer and more sophisticated techniques for both monitoring and modifying the synaptic activity inside human brains. And neuroscientists will continue to make progress in understanding how the brain works. Taken together, these trends point toward a startling conclusion: the direct transmission of mental content from brain to brain is no longer mere sci-fi fantasy. Some scientists now regard it as a tangible possibility, to be actively pursued as a research goal.

Now let us imagine, for the sake of argument, something considerably more advanced. You and I sit back to back while the technicians hook up the skullcaps and the computer. You close your eyes and call to mind the memory of the raspberry croissant you ate for breakfast this morning. You focus your attention on this memory, letting the skullcap pick up the neural patterns. After a moment, you press the handheld button for "Send."

What do I receive? The evidence available today suggests that a person's memories do not exist in the brain as discrete packages of information, akin to film clips stored digitally in a computer or on a DVD. This is a common misconception promoted by Hollywood sci-fi films.[7] Instead, neuroscientists and psychologists suspect that our memories are stored by the brain in a far more distributed fashion, with considerable redundancy in the neural patterns linked to specific experiences.[8] Particular memories probably do not exist as discrete entities at all, but rather as embedded elements in a vast webbing of associative connections, with distinct aspects of each experience linked to similar aspects of other experiences.[9]

This suggests that the task confronting the designer of a memory-sharing skullcap system would be very difficult indeed. The device must be able to zero in on the set of neural patterns associated with the specific content that a person is holding in conscious attention, while excluding all the concentric circles of associatively connected neural patterns that are indirectly linked to it. This challenge is daunting enough in itself. Then comes the even greater challenge of instilling a very similar neural pattern into the appropriate circuits of the brain on the receiving end.

Let us nonetheless assume for a moment that these obstacles were overcome at some point and that the brain-to-brain sharing system worked well: you transmit to me your memory of eating the croissant. How might

I experience it? It is likely that my secondhand experience of your memory of eating the croissant will differ significantly from your own firsthand experience. I would inevitably relate the incoming croissant sensations to my own storehouse of memories and associations—my own meanings connected with pastries, raspberries, breakfasts—and all this would give my resultant second-order experience a decidedly different "flavor" from the one felt by you. My experience of your memory would thus be recontextualized: it might bear great *similarities* to what you experienced, but it would by no means be identical.

Nevertheless, even with this very important caveat, the fact remains that a direct transmission of mental representations from one brain to another would constitute a real qualitative leap. Nothing remotely like it has ever occurred in the history of the natural world. A recontextualized memory of your experience eating the raspberry croissant, presenting itself in my mind as a direct transmission from you, would in some ways bring me closer to experiencing the world from your perspective than any other kind of sharing in human culture.

This would be a wonderful thing, for some purposes. I would be able to empathize with others in an extremely vivid way. Two lovers or friends could share their recollections and feelings with unprecedented potency. As a teacher, I could communicate with my students far more effectively, directly transmitting my own images and understandings, and receiving back their own reactions and distinctive ideas.[10] Two persons from different cultures could experience each other's worlds, allowing them to recognize the common ground that binds them together at a deeper level.[11]

All of which would presumably be peachy. Even so, troubling questions arise. Memories, and the feeling of those memories being uniquely *mine*, are central to our sense of who we are as individuals.[12] It is therefore hard to see how such an unmediated sharing of experiences would not challenge—and perhaps even destabilize—the boundaries of individual selfhood. Which memories are truly mine and which are yours? Even if I have taken care to tag the memories, upon receipt, as "not mine," it is likely that a prolonged and extensive sharing of intimate, rich, subjective experiences would eventually cause a blurring of the boundaries between me and you.[13]

In today's society, we generally consider such a meeting of the minds as a positive development, but that is precisely because the two minds in question remain clearly distinct from each other throughout the whole process.

We come together, we communicate and empathize with each other, we learn a lot about how the other person sees the world, and then we go our separate ways again, slightly altered by the encounter, but still the discrete individuals we were before. Precisely because the acts of communication and empathizing are limited, imperfect, and incomplete, they leave us with no doubt as to who is who.

The memory-sharing technology, by contrast, may prove to be (once again) too much of a good thing. In the story of Jed and Francesca, I depicted two individuals undergoing this kind of give-and-take transmission for the first time. But what would happen if they were to undertake this kind of intensive session two or three times per week over several months? Would it not be reasonable to expect that their sense of distinctness as individuals might begin to break down? In clinical psychology, the development and maintenance of a separate ego constitutes a fundamental aspect of mental health. When people fail to draw clear boundaries around "me" and "mine," the results are not pretty. They lose the central reference point around which to anchor their feelings, desires, fears, plans, motivations. The unity of their world comes apart, and they become profoundly dysfunctional.[14]

There is no reason, moreover, why this type of sharing would have to be limited to just two individuals sitting in the same room. The same technology that enabled this kind of person-to-person transmission would presumably also enable me to send and receive countless such experiences over the Web. I could upload the neural-encoding files relating to my own memories in the same manner as people do today with amateur videos on YouTube. And you, a total stranger on another continent, could then download my experiences and relive them, recontextualized within your own personhood and lifeworld. You and I could purchase, trade, or simply give away or freely take in just about any experience recorded by any connected human brain anywhere in the world.

The French Jesuit philosopher Pierre Teilhard de Chardin envisioned a phenomenon that he called the Noosphere—an entity akin to the geosphere and the biosphere, encompassing all human minds along with the totality of their collective interactions. According to Teilhard (who died in 1955), this sphere of mentation already exists in the contemporary world. It has been steadily growing in integration and intensity over the passing centuries, for that is the hidden destiny, the inner telos, that structures human history.[15] No doubt he would have been enthused about the Web and downright enthralled at the possibilities unleashed through a memory-sharing technology. I suspect that he would have regarded it—the gradual blurring of myriad individuals into a higher meta-identity—as a kind of secular miracle.

All of this is speculation, of course. Our brains may turn out to work in ways that render such retrieving and transmitting of mental contents simply impossible. Or, even if scientists conclude someday that it is possible in principle, it may still prove a mirage in practice: twenty-first-century bioengineers may never figure out how to make the requisite technology run properly.

The fact remains, nonetheless, that our society is currently investing huge amounts of money, resources, and ingenuity in developing precisely the kinds of scientific understanding and technological know-how that point in this general direction. It is therefore worth taking seriously the possibility that such devices might one day become available.

THIRTEEN

Virtual Reality and (Yawn) This Other Reality

As he flies his golden griffon through the blue skies to join with his army, Bob starts to feel so good. This. Is. Fun. Indeed this is the first moment of the day when Bob has actually felt alive.

—Edward Castronova, *Exodus to the Virtual World* (2007)[1]

- **VIGNETTE: *SHIVA AND THE CORGI***

"So where are you taking me?"

Tsien just grins and says, "You'll see."

He drives me downtown, to the Portal. "We have reservations for two at 11 a.m.," he says, as we pull into the parking spot.

I don't know what to say. We have a whole Saturday together, and he wants me to go into the Sphere with him?

He can see it on my face.

But he just smiles. Reaches over and lays his hand on my arm. "You'll see, Sumalee. It's not what you think."

Obviously, my face shows that I'm not convinced.

"Trust me," he says.

"OK, one hour." I look at him. "And afterwards, we're going down by the riverfront to Eastwood Park. And we'll buy one of those kites they sell for the kids. And we'll fly it together. A real kite, flying in the real wind."

He grins. "OK, sounds great."

We lie back in our gamer's harnesses, the skullcaps tight on our heads. The attendant bot makes the final adjustments, checking the connections to the mainframes. I've done this a couple times before with the kids I babysit, so I know the drill.

I glance over at Tsien, and he nods. I close my eyes.

A red lightning-bolt icon appears in my visual field. I select it.

A text box unfolds. "Broadband remote access requested. Authorize? (Yes) (No)."

I select yes and give my internal password.

The visual field goes black. A voice begins speaking inside me: male voice, deep and resonant. Probably the System AI. "Hello, Sumalee. You're going to need to choose an avatar before you come inside."

A menu of several dozen human figures appears. Stock images, ranging from fairy tale princesses to Nordic warrior gods.

Lord help us.

I scroll down and am relieved to see the options varying much more widely as the menu unfurls. Women, men, in all sorts of garbs, then animals, then mythical creatures, strange hybrid beings, aliens of all shapes and sizes.

A small red fox catches my eye, and I almost select it. It's beautiful and has a look of intelligence about it. But then it occurs to me that I might have to run around on all fours. And bark instead of talking. I hold off and keep scrolling. I wonder what Tsien has chosen.

Finally I find it: an Indian artist's representation of the god Shiva. Blue skin, four arms, long blazing black hair, a keenly mischievous expression, a dancing pose.

Well, I might as well do it in style.

"I'll take this one."

"Shiva the Transformer," says the AI.

The menu disappears, and suddenly I find myself standing under a bright yellow sky filled with black stars, out on an open desert plain. Kind of a hybrid of day and night. Barren land, a few large boulders here and there, saguaro cacti, a silhouette of snow-capped mountains in the distance. It's desolate and eerily beautiful.

I look down at my legs and torso: light blue skin, rippling pectoral muscles. A red silk cloth wrapped around my waist. I move my arms, and sure enough, four hands come round and hold themselves open before me. I wiggle the twenty fingers.

"Cool, huh?"

I turn, and there standing next to me is a small corgi dog.

"Tsien?"

"Yup," says the corgi.

I can't help myself. I laugh out loud.

"Look behind you, Sumalee," Tsien says softly.

I turn, and my eyes rise up toward the sky. There's a city floating there. Suspended in the air, astride a giant hovering rock about a half-kilometer up, an entire city. Sort of a mix between Renaissance and Mediterranean architecture—lots of arches, tile roofs, terraces, parks, tiny meandering streets, hanging plants, balconies with finely patterned iron railings.

And people. Now I can see people moving about up there. Vendors in the market squares, bicycles moving down cobbled streets. Old ladies carrying baskets on their heads.

I turn back and look at Tsien. Hard to tell what he's thinking: the corgi's facial expressions are limited.

"It's beautiful," I finally say. "Where are we?"

"This is the world I've created, Sumalee. This is what I wanted to share with you."

"You made this?"

"I did."

The corgi gestures with its paw, and a small black shape detaches itself from the floating city and glides rapidly down toward us. It's an aircraft somewhat like a helicopter, only it has no rotor blades. And it's perfectly soundless as it slows and gently touches down on the soft sand.

Tsien and I climb in.

Silence as we glide upward. Like an elevator.

The view spreads out in all directions, the desert expanse, the mountains. In the far distance, I see another floating rock with a city on it. Then I notice another, way off to the left. A big one. I think I can dimly make out an oceanscape over there.

I point at the distant city and glance at Tsien.

"That's Portland," he explains. "Built by a friend of mine. My city's people do a lot of trade with his city's people."

Aha. Here's where the blood and mayhem come in.

"Do they ever go to war with each other?"

"This is a world without war, Sumalee."

I look across at him. He's built a utopia, here in the Sphere?

"What if your people feel like they're getting ripped off by the deals they make with Portland's people?"

The corgi gestures with its paw. "They negotiate."

"What if the negotiations fail?"

"They negotiate more."

We glide up through a huge arch into a hangar bay, and the aircraft smoothly docks itself.

People moving about on the sidewalks. Small electric vehicles on the cobbled streets. An old man selling huge bunches of flowers near a walled garden.

"Are . . . are they all avatars of different players? Each of those people I see over there?"

"No, they're controlled by the AI that runs the world-building software."

"So the AI runs the place?"

"No, I do. I'm the one who designed it. I'm the one who sets the behavior parameters."

We climb out of the copter and stand on the platform, gazing up at the rising shapes of the city above us. Streets, bustle of activity, rooftops with their red tiles. For some reason, it reminds me of Lisbon, the old neighborhoods along the hillsides.

"How long have you been working on this?"

"About three years."

"Three years!"

"I do most of the rendering and programming from home, then load it into the master database via the Web." Slight pause. "I can't afford to actually come in here very often."

"Doesn't it take a lot of your spare time? Like, evenings and weekends?"

"Yeah, it does. At least, until three weeks ago when I met you. I haven't gotten much done on it since then."

I smile. Especially after the second date, we've been spending a lot of time comming each other and talking on the phone.

I reach over and touch a nearby trellis vine with one of my right hands, the blue fingers running along the foliage. I can feel the individual veins in each leaf. Every little detail. It's a labor of love.

"It really is beautiful."

The corgi looks out into the nearby streets. "For me, it's the satisfaction of setting up a whole world, tending to its needs, making sure it runs well. Which is not easy, it turns out. Doctors, trash collectors, musicians. Parking fines, stray dogs. Trade unions, bowling allies, burglaries." He gestures with his paw. "Things are always surprising me in this world. There's this constant question: Am I going to be able to keep it going, or is the whole thing just going to implode on me?"

I gaze out over the bustle of people. A young woman over there has stopped by the flower vendor and purchased a spectacular bouquet of red and yellow zinnias.

Tsien's world.

"What's your city's name?"

The corgi looks up at me, then away again.

"I never gave it a name until two weeks ago. I just called it 'The City.'"

"Why?"

"Because I couldn't think of a name that matched how special it was to me."

Silence.

"So what'd you name it?"

"You know what I named it."

. . .

Around the globe today, approximately 660 million persons—about a tenth of humankind—partake regularly in video or online games, spending (on average) thirteen hours per week in the virtual world of their choice.[2] The sheer scale of this phenomenon and the steep growth curve of its expansion, year after year, have led some scholars to speak of a "mass exodus" into virtual reality.[3]

Today's virtual reality technologies fall into three main categories: expensive laboratory systems that provide fully immersive visual experiences via a headset and a complex interactive console; home devices that engage the senses via a TV or computer screen and simple handheld controls; and emerging new home devices such as the Oculus Rift that offer immersive experiences at more affordable prices than the laboratory systems.[4] People who use the lab-based immersive systems report an experience that feels very much like the real thing. For instance, if the simulation entails walking across a narrow wooden plank spanning a deep chasm, they feel genuine fear, their heart rate goes up, they sweat profusely, and they have a surge of elation and relief when they reach the other side—all this while they know perfectly well that they are standing safely on the floor in a laboratory, wearing a headset. The human mind, apparently, toggles quite easily into the immersion experience, reading the sensory signals as real, and responding with the full range of appropriate somatic, emotional, and cognitive reactions.[5]

This psychological effect goes a long way toward explaining why video games and other such interactive-fantasy technologies have burgeoned into billion-dollar industries in recent years.[6] People who play these games experience a full panoply of strong emotions while taking part in them—real fear, dismay, joy, comradeship, humiliation, triumph. At one level, the source of the emotions may be accurately describable by an outsider as "simulated," but to the person experiencing them, this distinction is irrelevant: the fun is real, because it *feels like* fun.[7]

Three trends will probably mark the development of these "synthetic experiences" over coming decades.

1. *Lifelike realism.* The simulation technologies will become increasingly immersive and convincing, to the point where it becomes hard to distinguish between virtual adventures and the things a person does and feels in real life.[8] Gamers will have the option of scanning their own bodies,

faces, and behavioral quirks in minute detail, thus begetting avatars that look, move, and feel like faithful copies of their real selves.[9]

2. *Seamless access.* The rising sophistication of bioelectronic enhancements will probably revolutionize the way we interact with virtual reality technologies. Our movements, reactions, and intentions will be translated far more fluidly into subtle behaviors in the virtual world. This increase in transparency, in turn, will further ramp up the realism and emotional impact of the simulation.[10]

3. *Blurred boundaries.* Primary reality—the physical world of our everyday lives—will itself come to be partially overlaid by elements of the virtual: the technical term for this phenomenon is "augmented reality."[11] If a person wears the appropriate devices—something like an advanced version of today's Google Glass—the physical objects she encounters will "talk back" to her and interact with her in powerful ways. She walks down the sidewalk, for example, and the street comes alive with maps, directions, and practical advice projected into her visual field. In the supermarket, she reaches for a box of cereal; the box flashes nutritional information, as well as reviews of the product by other consumers.[12] It amounts to a sort of hybridization of the physical and virtual, and it is a trend that we can expect to continue unabated.[13]

VIRTUAL BENEFITS, REAL DANGERS

Every year, humans around the world spend about 442 billion hours of their time frantically clicking away in interactive online gaming.[14] Isn't this all a colossal waste of time, a mass indulgence in collective escapism? Definitely not, argues the researcher and game developer Jane McGonigal: online gaming is not at all about frivolousness and mere play. Au contraire, the hallmark of a good game design is that it engages the participant in a particular kind of intensive *work*.[15] The best games, she maintains, provide constant feedback, tangibly showing the player's progress through a rising array of challenges and levels. They offer the risk of failure and a plausible chance of success. And they are framed within a social context where one's achievements and exploits will be recognized and appreciated by other people. Perhaps most importantly, the best games take place as quests or missions: they operate within a storyline that imparts a strong sense of self-transcendent purpose to one's decisions and actions.

But it's still just a fantasy, one might object. You're not really a hero, once you log off. Perhaps, argues McGonigal, but that's not the way it feels.

To millions of people who participate regularly in these games, their virtual adventures are more rewarding than many of the real-life pursuits they find themselves compelled to undertake over and over again in the humdrum routine of their daily lives.[16]

This does not mean, however, that McGonigal is urging us all to pack our bags and take the plunge together into online game land. Quite the contrary: she argues that we should use the practical insights gained from the study of gaming to improve primary reality itself. If we understand what makes a well-designed game so powerfully appealing, this can offer us hints for modifying our everyday lives accordingly, thereby rendering our jobs, classrooms, relationships, and routine chores more enjoyable and fulfilling.[17]

One of the most fascinating aspects of virtual reality technologies lies in what the researchers Jim Blascovich and Jeremy Bailenson call the "Proteus effect."[18] Your simulated experiences, it turns out, can profoundly influence your self-perception and choices back here in primary reality. In a series of experiments at Stanford University and the University of Chicago, for example, male and female subjects were given avatars that were either taller or shorter than the subject's real-life body.[19] After "inhabiting" their avatars for a while online, the test subjects were presented with virtual situations that required negotiation and compromise with other players. Almost without exception, the subjects with the taller avatars showed markedly greater self-confidence and dominance than those with short avatars. This was not particularly surprising, since psychologists have long established that taller individuals in the real world tend to enjoy greater advantages (and confidence) than short people in interpersonal interactions.[20]

But then came the kicker: after removing their headsets and exiting the simulation, the subjects went through another round of similar tests, this time in the laboratory setting. Regardless of the height of the subject's real-life body, and regardless of whether they were male or female in real life, those who had temporarily used taller avatars continued to display significantly more assertiveness than those with diminutive avatars. Like the Greek god Proteus, their ability to change body shapes (virtually) brought about a concomitant shift in real-life persona and character.

This could be a good or a bad thing. At one level, it could allow people to take a short vacation from their real-life selves, experiencing the world in radically new and liberating ways. An Iranian Muslim could temporarily experience the world from the perspective of an Israeli Jew—and vice versa. A rich person in Hong Kong could spend a day among street urchins from the favelas of Rio de Janeiro. Our virtual experiences could serve as a potent tool for broadening our horizons.[21]

Other researchers such as Sherry Turkle and Jaron Lanier, however, have taken a less enthusiastic view. While acknowledging the extraordinary potential of the online world, they feel that the way we use these technologies *in practice* has resulted in a subtle degradation of our social relationships. The problem they identify is twofold.[22] First, too many persons tend to adopt these technologies not as a complement to their real-life experiences, but as a substitute for them. They gradually lose themselves to it, and their real-life relationships and pursuits drift imperceptibly toward atrophy. Turkle's rueful conclusion: "we expect more from technology and less from each other."[23]

The second main problem with these technologies is that they inevitably oversimplify the world. A digital environment cannot help but be a stylized construct, a schematic representation whose creators have made all kinds of choices about what to exclude from the scene's features and protagonists.[24] According to the researcher Edward Castronova, it is this very simplicity that gives virtual worlds their powerful appeal. Consider the story of Bob, depicted in the epigraph for this chapter, as he transforms himself into the heroic warrior Abelaard, combating Azengoth, the Demon of the Underworld. Castronova concludes that online games are providing people like Bob with a rich source of purpose and personal validation, precisely *because* they offer clear-cut narratives in which the lines between good and evil are so starkly drawn. In the post-Nietzschean era, with its rampant relativism and squalid materialism, online games provide ordinary folks with a badly needed context of myth and idealism, imparting moral orientation to lives that have otherwise come to seem empty and pointless.[25]

Castronova makes a perceptive point here, but in the end, his spirited defense of virtual reality fails to persuade. This type of stylized moral script may feel good to its practitioners in the short run, but it ultimately leads to a stunted vision of life. The real world is fraught with ambiguity, complexity, and lack of resolution, but it is precisely these features that challenge us humans most intensely, forcing us to struggle for ever higher forms of understanding and wisdom.[26] To the extent that you spend your time on this earth chasing two-dimensional characters like Azengoth, you are settling for a dumbed-down and infantile range of challenges.

Over the coming decades, therefore, we would do best to regard the simulated worlds as a potent but insidious invention. Handled judiciously, those technologies can plunge us into challenging situations that stretch our skills and deepen our character. But they also have a tendency to seduce us, leading us away from the seemingly mundane relationships and commitments that ground our lives in primary reality. If we aren't careful, they can imperceptibly disconnect us from the things that matter most.

FOURTEEN

Till Death Do Us No Longer Part

Implications of Extremely Long Health Spans

Yeah, well. The Dude abides.

—Jeff Lebowski (Jeff Bridges)
in *The Big Lebowski* (1998)[1]

■ **VIGNETTE:** *MEETING AN ANCESTOR*

I lie in bed at the Westminster Manor Hotel, thinking about my great-great-great-grandfather. His name is John R. G. Winslow. Brits have two middle names, I found out.

I met him this morning for the first time. He just celebrated his 141st birthday last week. He looks about four or five years older than I do. I'd say about thirty-five, thirty-six. The juve treatments have really worked in his case. He looks . . . amazing. Tall, muscular build, a thin mustache and goatee, wavy red hair. Maybe thirty-seven at most.

We gave each other a big hug when he came to fetch me at the hotel this morning. Standing in the foyer, sizing each other up. He held me by the shoulders, his eyes bright with excitement.

"Well, granddaughter," he says, "there's definitely a touch of the Winslow still in you. No doubt about it. I can see it plain as day!"

He turns me toward the gilt-edged mirror on the wall nearby. My petite Asian features and curled black bangs, John's figure looming beside me, broad grin.

I roll over in the bed, hugging the pillow. This is wrong, I keep saying to myself. This is crazy.

He whisks me out to his car, the doorman bot holding the car door for me as I get in.

"Have you had breakfast, Keiko?" John asks.

I shake my head. "Forgot to bring the jet-lag pills. I was up half the night. Slept till ten."

He nods. "Perfect. I know just the place then."

He tells the car, "Marks & Spencer."

"Main store or one of the branches, sir?" queries the car.

"Have I ever gone anywhere but the main store?" asks John, irritated.

"Just checking, sir." The car pulls swiftly into the traffic, the flow of vehicles parting sinuously to let us in.

I'd thought Marks & Spencer was just another department store, silly me. They have the most miraculous little white-bread sandwiches. Each about six centimeters across. Dozens of varieties to choose from. Hundreds. I didn't think I was hungry, but John made me pile my plate high anyway. Then we take our trays out to this rooftop terrace filled with potted plants and a trellis with creepers, matronly ladies in their floral hats having tea and crumpets, the background noise of London emanating from the streets below.

Turns out we like all the same kinds: anchovy and tomato; cucumber and soy fritter; mushrooms and dill cream. I don't mean most of the same kinds. I mean *all* the same kinds.

Funny thing was, it was effortless. I'd been nervous the whole flight over from Osaka. What's this going to be like? What if we don't have anything to say to each other, after the first few polite exchanges?

It's nothing like that. The exact opposite. We talk fast. We finish each other's sentences. We jump from subject to subject. We laugh like mad. And it's completely relaxed, easy, uncomplicated.

Good thing I wore the shorts and blouse instead of the calico dress. After brunch we rent a couple bicycles, ride down to the Thames, cycle along the riverfront. We stop at a bench near some elm trees and sit and watch the barges going to and fro, the tourist boats, the yachts. For a while we sit in silence, and once again I'm astonished: the silence itself feels natural, peaceful. It's a silence that bridges two people instead of dividing them.

I look at him and say, "Do you think it's the genes? Is that why we're getting on so well?"

He shrugs. "Five generations. Pretty distant genes at this point."

I press him on it. "You're the one who had a whole career in biology before you became a journalist."

He chuckles, gazes out over the water. "And a whole slew of careers before that."

I ask him why he hasn't remarried. The way he explains it, it makes sense. Two long marriages are enough for a while. Thirty years with Mona, from whom my own family line was born; fifty-five with Suzanne. He grins at me. "It was wonderful, both times. A gift. I'm so grateful for it." Then he shrugs again. "Now I'm just experimenting with being on my own again."

"So what have you been doing?"

"Whatever I please. I've got enough money saved up so I can make up my own rules, day by day."

"What sorts of things?"

"Well, I spent three years at a monastery in Burma doing Buddhist meditation, for instance."

I stare at him. "What was it like?"

"Slow."

I wait for him to go on.

"People tend to believe the juve treatments and antiaging pills are the way to expand your time on this earth. But there's a simpler way. You just learn how to pay attention to each passing moment. Remember the William Blake poem 'To See a World in a Grain of Sand'?"

I nod, vaguely recalling a poem along those lines.

"Just being present for the minutes that come to you. No need for fancy technologies."

Silence.

We watch the boats on the river.

Then he turns toward me. "How about you, Keiko? Your cousins in South Africa told me you've never been married."

I laugh. The South Africa branch of the family. Ardent non-mods. No juves, no pills, no epigenetic design, no bioelectronic implants, no upgrades or boosts. Just tulips. The whole extended family, all living on that rugged farm in the Transvaal with its hectares of greenhouses. Couldn't care less about the rest of the world running circles around them.

"So you've been in touch with them, have you?"

He nods. "Quite a bunch, aren't they?"

We both laugh.

On impulse I reach over and touch his arm. I get up and grab my bike. Race you, I say, and I'm off down the riverfront.

Pretty soon I hear him behind me. He glides past, twice my speed, effortlessly.

Nobody told me he's on the All-England amateur team.

My chronological age is thirty-one. I look and feel pretty good, thanks to the juves. But I'm nowhere near being in the shape he's in.

I finally catch up with him, gasping, at the end of the tree-lined causeway. He's got his bike tipped up against a railing, adjusting something on the pedals.

"How about a little visit to the Tate Modern?" he asks, without turning around.

■   ■   ■

I sit up in the bed, turn on the light. The bedside clock reads 2:33. The middle of the night. I'm sitting here in London, England, thinking about a man who's a blood relative. Thinking about him in the wrong ways.

I have to stop this.

After a while staring at the walls, the faded paisley imprint of the wallpaper, the stodgy hotel furniture, I give up and turn off the light and lie back in the bed.

I could take a mood pill, of course, and feel fine within three or four minutes. Peaceful, clear, contented.

But this is too important for that kind of composure. I need to get a grip on myself and think this thing through.

What would he say if he knew I was lying here thinking of him?

Thinking of him in this way?

Is it really that wrong? He said it himself: five generations. Things get pretty diluted, after all that mixing.

How many degrees of relation have to separate you from someone, in order for it not to be incest?

Before going to bed I went online and read up on it. The laws are a mess, different in every country. But the basic standard seems to be: four generations at least.

So legally we'd be OK. Just barely.

But what about these feelings? What would he think if I revealed them to him?

It would help if he weren't so damn good-looking.

■   ■   ■

The only two certain things in life, so the saying goes, are death and taxes. It turns out, however, that taxes may prove to be the more rock-solid of the pair. If you've been investing lately in the stock of funeral homes, you might want to reconsider.

In chapter 2 I described some of the reasons why our bodies deteriorate as we get older.[2] Scientists now understand aging as a process involving a wide array of interacting cellular and metabolic mechanisms, and they are zeroing in on strategies for altering those mechanisms, one by one, in ways that could cumulatively have a dramatic impact on the average human health span.[3] In this chapter we consider what would be the results if they succeed.

In order to keep the discussion coherent, I will make the following assumptions. First, I will assume that medical advances over coming decades will gradually lead to a life expectancy of about 160 years for most humans

in prosperous countries—roughly double the figure of today.[4] I will further
postulate that these advances will allow us to live active, healthy, mentally
vigorous lives until the age of about 145 or 150, at which point the deterioration of our bodies will be swift and implacable. This latter concept
is referred to by aging specialists with the charming term, "compressed
morbidity."[5] The idea here is that few persons would want to live on for
decades, well past the age of 100, under conditions of deepening physical
and mental debility. What people want is a greater health span, not just a
longer life span.

These assumptions are, of course, arbitrary at one level, but they are
more or less in line with the conjectures that some scientific experts make
when asked how far the human health span could conceivably be stretched
over the coming century.[6] It is intriguing to play these possibilities out and
see where they lead.

ECOLOGICAL CONSEQUENCES

If people were to live so much longer, and new babies were still being born
every day, how could the earth sustain all those accumulating generations
of humans? Wouldn't life become a wretched scramble among the seething
billions for ever scarcer resources? The answer, it turns out, is not as grim
as one might think. Everything hinges on two main variables: fertility rates
and the earth's carrying capacity.[7]

Global fertility rates have been going steadily down over the past century: they currently stand at 2.55 children per woman and are projected
by the UN to continue declining to about 2.02 children per woman by the
year 2050—roughly the same fertility level that characterizes the United
States today.[8] The demographers Leonid Gavrilov and Natalia Gavrilova,
at the University of Chicago's Center on Aging, have run extensive computer simulations of future population growth. To their astonishment, they
discovered that once fertility rates drop below the magic number of 2.0
children per woman, a population can live longer and longer without necessarily experiencing massive overcrowding.[9] Their research suggests that,
if today's entire global population were given "longevity pills" resulting in
health spans up to 160 years, and if they maintained the current average
fertility rate of 2.55, the long-term result could indeed prove catastrophic.
The seven billion copulating humans would gradually swamp the planet.
But if one assumes that human longevity were only increasing incrementally,
decade by decade, while fertility levels were simultaneously continuing their

current overall decline, the outlook becomes much less depressing. Once the global fertility rate drops below 2.0, a human life expectancy of 160 years would not necessarily have to result in massive overcrowding and ecological collapse. (Indeed, as long as *everybody* was still eventually dying at some point around the age of 160, then a global fertility rate below 2.0 would result, over time, in a gradual decline of the total population.)

Which brings me to my second point: what counts most is not the number itself of mouths to feed, but rather how we go about feeding those mouths. Today's average American citizen has an "ecological footprint" that is roughly twenty times larger than that of the average Mozambican citizen.[10] Thus, if all the earth's seven billion human beings consumed resources at the same rate as the average American, the total ecological impact would be dramatically different than if they consumed resources at the rate of the average Mozambican. The question then becomes one of politics and ethics rather than mathematics or ecology: the planet could support relatively higher population levels if humans found ways to restrain their lifestyle needs and to adopt new technologies that allow them to live in a more ecologically sustainable fashion.[11]

All this is a lot easier said than done, of course. But if, over the coming century, fertility rates gradually come down below 2.0, while sustainable economic practices become widespread, then a doubled human health span need not spell disaster. On the contrary, it could comprise one element within a broader profile of spreading prosperity, health, opportunity, and ecological balance.[12]

THE MULTIGENERATIONAL FAMILY

Over the coming century, as health spans extend out toward the 160-year mark, the institutions of marriage and family will probably be among the first to feel the pressure.[13] What will it be like, under these new circumstances, to stand before each other at the altar and say "I do"? To some individuals, no doubt, the prospect of spending 130 years with the same life partner may appear as a blessing. (I am pleased to count myself as a member of this category, dear.) For many others, however, this will not be the case. Like my character John Winslow, they may find themselves entering into a series of marriages over their lifetimes.[14]

What kinds of families would emerge among such long-lived people? Let us imagine one fictional family, the tight-knit Brown/Smith/Joneses of Dayton, Ohio.

THE BROWN/SMITH/JONES CLAN

Maynard Brown (137) ---- ┬ ---- Tricia Brown (132) [divorced]

Silvia Brown (105) ---- ┬ ---- Hank Smith (105)

Archibald Jones (105) --- ┬ --- Mary Smith-Brown (77)

Gina Thompson (66) --- ┬ --- Fred (51)   Alice (11)

Audrey (22) --- ┬ --- Maria (94)

Gerald (3) [adopted]

What we see here are most of the traditional phases of life, dramatically scrambled and intermingled. One pictures the whole clan seated round the table at Thanksgiving dinner: eleven individuals ranging in age from 137 to 3—all the adults relatively youthful looking, thanks to rejuvenation treatments (though still varying considerably in appearance). Audrey is twice the age of her aunt Alice. Little Gerald has two mothers, one of whom (Maria) is 17 years older than his great-grandmother (Mary). Hank Smith and Silvia Brown are the same age as their daughter's husband. Maynard Brown carves the soy turkey and beams at his new great-great-great-grandson Gerald, who is seated next to him in his high chair. Maynard is thinking: "Just four more years before we can get him seriously started on tennis."

Compared with the nuclear families of today, such a "post-nuclear" family would have several advantages and disadvantages. On the positive side, the older members would be in a position to extend considerable assistance to the younger ones in all manner of ways. They will have had the opportunity to amass substantial wealth over the years (more on this below) and will be able to offer not only direct financial help, but also valuable advice and a broad network of personal contacts to their descendants as they get started with their own careers. Families will therefore constitute a repository of accumulated financial and experiential capital that far exceeds the levels witnessed today.[15]

At a more personal level, individuals will possess unprecedented opportunities for intergenerational encounters and bonding: a young person like the twenty-two-year-old Audrey will be able to know firsthand what her grandparents, great-grandparents, and great-great-grandparents are like, since she will frequently be interacting with these ancestral figures. The impact on one's sense of "family roots" could prove profound. For women, moreover, the new patterns of longevity will offer much more flexibility in the timing of fertility and rearing of children. For a woman whose biological aging processes have been greatly slowed, it is probable that menopause can also be significantly postponed, thereby allowing her the option of having children later in life, under conditions of her own choosing.[16]

On the negative side, the huge gaps in age between siblings and between spouses could pose new kinds of problems. Someone like Fred, who is fifty-one, will have missed out on sharing the myriad experiences (good and bad) of growing up together with his sister Alice, who is forty years his junior. Although she is technically his sister, he will tend to relate to her more like a father or uncle than a brother.[17] A similar type of experiential disparity will tend to separate two individuals—like Audrey and Maria—who fall in love across the span of many decades in age difference. Such individuals may *look* relatively close in chronological years (thanks to rejuvenation treatments), but beneath the surface, one partner will be carrying around a vastly different set of memories, knowledge, cultural associations, and generational sensibilities. When the initial rush of infatuation wears off, many such cross-generational loves may deteriorate or break down altogether.

CAREERS AND ECONOMIC LIFE

The concept itself of "career" will probably undergo a drastic transformation.[18] Where today we might switch from being a news reporter to, say, writing advertisement copy or editing manuscripts, the career shifts of the late twenty-first century could go from lawyer to biologist, from farmer to airline pilot. After forty years as a doctor, I may simply decide I want to try something *really* different and start exploring my options in astronomy or massage therapy. With a 160-year health span, I will have all the time I need to make this happen.

Such a model would entail multiple career phases, punctuated by periods of rest, reeducation, and retooling.[19] You work at your firm for thirty or forty years, taking care to pay a portion of your salary into a special reeducation annuity.[20] When the time is ripe, you throw a farewell party and say goodbye to your colleagues. After a breathing period, you draw down your annuity over three or four years to finance your return to the university setting—whether as a resident student or, more likely, through telecommuting and online courses. You take your classes, learn all kinds of cool new stuff. (Optional: football games, beer pong, Boola-boola.) Then you graduate and go back out on the job market, eager to ply your next trade.[21]

Some individuals, however, will no doubt reject this paradigm. After diligently working their way up to the top of their firm over several decades, they may refuse to just let it all go and start over. They might even be inclined to hang on for a whole century, seeking to remain at the company's helm from age 50 to 150. Even if these centenarian bosses managed somehow to remain fresh and innovative in their leadership styles, their presence over decades and decades would leave little room for younger employees to move up in the company. Indeed, this problem would presumably spread across the entire economy, as the upper reaches of most enterprises would gradually come to be saturated with very old (but still sharp and youthful-looking) men and women. The only solution, in the end, may be to mandate some form of "term limits" for senior management, beyond which one would be strongly pressured to move on—either into retirement or into a new career somewhere else.[22]

Retirement itself will of course be quite different from what it is today. You could retire at age ninety or a hundred with a good half century of youthful vitality still ahead—and with a significant pot of cash, if you played your cards right. Here the mathematics of compound interest comes into play. If you set aside $500 per month and put it in the bank at 5 percent interest, starting at age twenty-five, then by the time you reach age sixty-five, you will have $766,000. But if you keep doing this for an additional forty years, and retire at age 105, your account will have compounded itself up to a whopping $6.4 million. Keep at it another twenty years, to age 125, and you've got $17.5 million.

This would presumably buy a lot of golf and cruises. Nevertheless, I suspect many retirees, finding themselves still full of vim, would reject any notion of retirement as a period dominated by leisure or restful pastimes. If they so wished, they could apply their half century of freedom to tackling a variety of personal and societal challenges, from self-education to travel, from mentoring youngsters to volunteering their accumulated expertise for all manner of urgent projects around the globe. Taken collectively, their impact on world affairs could be immense.

What about boredom? Some writers on extreme longevity believe this will pose a major problem, and that humans will eventually tire of repeating the same basic deeds and functions, decade after decade: flossing your teeth, preparing meals, taking the dog for a walk, and so on.[23] This is a valid concern, and I expect that our long-lived descendants will need to find ways to inject a broader sense of purpose and meaning into their daily activities, harkening to the value of Transcendence described in chapter 6. If they learn how to shake up their lives from time to time, taking on new creative pursuits and new practical challenges, and orienting their energies toward the countless ongoing problems that humankind will still undoubtedly face, then the afflictions of monotony and world weariness might be effectively kept at bay.

FIFTEEN

New Sounds for the Old Guitar

Sex, Food, Privacy, the Arts, Warfare

Police inspector: "You are playing God!"
Dr. Hfuhruhurr: "*Somebody* has to!"

—Dr. Hfuhruhurr (Steve Martin)
in *The Man with Two Brains* (1983)[1]

- **VIGNETTE: *ENCOUNTER***

The man and his machine came down the sidewalk. The machine, a somewhat outdated household bot, less than a meter tall, rolled half a pace behind. The man was in his forties, tall, lean, his short-cropped brown hair tousled and unkempt.

"Care to station here today, sir?" asked a parking meter. The man rolled his eyes and glanced at his machine companion.

"Griswold, why does it think I have a car?"

"Beats me," said the machine. They walked on.

"Got a special on eight-hour stays, sir," said the next parking meter.

Jesus, the man thought. Without looking aside, he said, "Tell 'em to bugger off, will ya?"

The machine opened its web link and got the codes of all the meters on the street, throwing in the billboards for good measure. *No Solicitations*, it signaled.

The street became hushed.

They came around the corner by Walgreens. A woman was standing beside the bicycle rack, her handbag propped against the wall. She glanced their way, then acted as if she hadn't seen them.

Wow, the man thought. I haven't seen her around here before.

She was beautiful.

Say something to her, commed the machine in private mode.

The man ignored this. His stress was going up, the machine noted.

C'mon, the machine urged.

She glanced at them again as they approached, then looked away.

The man hesitated. She really was beautiful.

Time was running out. Now or never.

He opened a microwave link on a clear channel and somewhat gingerly pointed it at her. She smiled slightly, but didn't look up.

He beamed her again, a bit stronger but not too much. He was only a few paces away now.

Still nothing. She might have been smiling, but he couldn't tell for sure from this angle.

One more try, then he'd leave her be. He modulated the microwaves subtly.

This time, to his astonishment, she opened up, though she still kept her eyes on the pavement.

They exchanged signal as he passed.

Just a fleeting couple seconds, wordless, intense.

Back and forth, to and fro, a crescendo.

Bursting between them.

She laughed. A beautiful clear laugh. Her face flushed now.

He smiled back, kept walking, the pharmacy doors opening with a swoosh. Didn't want to be obnoxious about this.

He was grinning, his heart beating fast, as he strode into the store and down the aisle. Inadvertently he glanced aside at the bot rolling next to him.

"Slut," said the machine.

He grinned even more. "Mind your own damn business," he said happily.

■   ■   ■

As the decades go by, most people are likely to adopt multiple kinds of devices, drugs, and genetic modifications concurrently, in all manner of combinations. Your epigenetic adjustments will interact with the performance-boosting drugs you take; these two in turn will interact with your brain-machine interface, which will simultaneously be interacting dynamically with the networked implants and devices of your friends and family. All these technologies will be intermingling, interfering with each other, and cross-fertilizing each other, in spirals of mutual causation. Some of the effects produced by these synergizing causal forces will be anticipated and well managed; others will result in emergent phenomena that, like the weather, are resistant to full control. There will be surprises. We should expect them.

I want to briefly consider five examples of these kinds of synergy, speculating about how they may play out in actual practice.

SEX

What might sexual behavior look like in a population of bioenhanced humans?[2] One can imagine, for example: –

- A pill that temporarily bypasses the rational faculties to enhance pure feeling, allowing deeper communion between two individuals (without the stupor or loss of self-control that characterizes the alcoholic beverages of today).
- Somatic gene packs or epigenetic modifications that boost one's sense of touch or, alternatively, that generate new erogenous zones (presumably requiring substantial additions to the Kama Sutra).
- Pills to augment various aspects of sexual prowess and performance, for both men and women.
- Bioelectronic devices (like the one described in this chapter's vignette) that offer entirely new dimensions of erotic contact.
- The ability to switch gender and sexual roles more easily than today, or to meld different aspects of gender and sexual roles in new combinations.
- The possibility of intercourse among more than two partners simultaneously, via brain-machine interface technology—resulting in "crowd orgasms."[3]
- Brain-machine interfaces that allow individuals to bypass the body entirely, engaging in intercourse directly from mind to mind.

Machines will no doubt come to play a considerably greater role in human sexuality than they do today. The combination of virtual reality technologies and brain-machine interfaces will probably yield extremely lifelike forms of interactive pornography, downloadable off the Web.[4] As robotics and artificial intelligence become increasingly sophisticated, it is likely that advanced android sex partners will come to replace the various forms of inanimate sex dolls and toys in use today. The writer David Levy, in his book *Love and Sex with Robots*, argues that humans will eventually come to develop full-fledged emotional relationships with such androids, since these robotic companions will have become sufficiently complex in their mimicry of human behaviors that people will spontaneously experience strong feelings of affection and even passion for them.[5] Levy's argument ultimately fails to persuade, however. It is one thing to develop a vague sense of affection for your old Buick, which has accompanied you through so many journeys and experiences over the years. It is quite another to experience a feeling of highly charged romantic attraction for a "simulated human"—no matter how artfully constructed the android may be. Most people would recoil at the prospect of investing genuine emotional intimacy into interactions with an entity that they know full well to be a sophisticated machine.[6]

Another watershed in the history of human sexuality would be the development of ectogenesis—the technology of artificial wombs.[7] Here, too, our society has already advanced farther down the road than most people realize. Human conception has been taking place outside the womb since the 1970s (through in vitro fertilization), and steady advances in neonatal medicine have allowed prematurely born babies to survive at earlier and earlier stages of growth. The current record stands at twenty-one weeks, and a birth weight of one pound, six ounces (resulting eventually in a healthy child).[8] To be sure, this remains quite far from full-blown ectogenesis. The biology of gestation is staggeringly complex, and it is possible that many decades will go by before a human being can be both conceived and carried to term within an artificial apparatus. Nevertheless, if such a technology were to become widely available, yielding healthy babies without the need for a woman's uterus, the societal implications would prove far-reaching.

Controversies over abortion would shift dramatically, because a woman who accidentally got pregnant would no longer have to end her fetus's life in order to terminate her pregnancy.[9] Instead, she could have the fetus surgically removed at an early stage and placed within an artificial womb to be carried healthily to term. The newborn infant could then be put up for adoption. Such a technological option would significantly increase women's reproductive choices and control over their own bodies.

Motherhood would also undergo a profound change. A woman's role in begetting a child would become similar to that of a man: she would provide one of the two gametes out of which the conceptus was created. (I am assuming here that egg and sperm would still come from a woman and man, though it is possible that synthetic gametes might also one day be invented.) From that point forward—at least, from a biological standpoint—males and females would play equivalent roles in parenting. People who so desired could bypass the traditional biological alignment of male and female in reproduction: two or three or more men could join together to coparent a child, as could two or three or more women—or any combination thereof.[10]

With human reproduction thus wholly separated from sexual behavior and gender roles, it is likely that the societal division of labor between the sexes could become truly equal for the first time in history. Women would also be free to modify their bodies in more radical ways than ever, since they would no longer need to take into account the basic anatomical requirements for childbearing and breast-feeding. Over time, this could result in a significant shift in the physical appearance and performance profile of women.[11] *Vive la différence . . . qui n'existe plus.*

FOOD

If you wish to enhance the way your body acquires nutrients and energy, two pathways are available: you can modify the food you eat, or you can alter your metabolism. It is likely that both strategies will be exuberantly pursued by humans over the coming decades.

New kinds of engineered foods will appear in the aisles of the supermarket. Scientists today are already experimenting with lab-cultured meats, and it is reasonable to expect that your T-bone steak, in the second half of this century, will no longer come from a slaughtered animal.[12] It will instead be grown in vitro, preformed in the shape of your favorite cut. It will probably taste delicious and deliver a broad profile of healthy nutrients. An ongoing collaboration will develop between master chefs and bioengineers, ultimately resulting in innovative hybrids such as banana-cilantro-kumquats, or cultured alligator/wildebeest osso buco. New designer spices will add piquancy to the day's cuisine. Self-preparing meals will be available, some of which will be engineered to have an extremely long shelf life. For the more adventuresome foodies, there could perhaps be meats and vegetables derived from the DNA of extinct species: pterodactyl wing soup or salad made with delicate azolla ferns from the Eocene period.

In the not-too-distant future, pharmaceutical companies will begin selling pills that partially reprogram your metabolism, allowing you to eat a towering ice cream sundae without getting fat or endangering your health.[13] Other pills will perhaps simulate the physiochemical effects of strenuous exercise, thereby allowing people to spend more time as couch potatoes if they so desire (though I suspect that some form of regular workout will continue to be required for maintaining basic health). Epigenetic or genetic interventions could conceivably augment the way your taste buds operate or (more directly) modulate the way your brain processes signals sent from the tongue and nose to the gustatory cortex. The result could be new levels of intensity, or subtlety, in the perception of how various foods and drinks taste.

Some forms of enhancement research, probably funded by the military, will explore the outer limits of how a human being can derive vital energy and nourishment in novel ways.[14] Food scientists in the employ of the US Department of Defense have already been working in recent years to develop the equivalent of "food pills"—compact and durable ways to deliver essential nutrition to soldiers under battlefield conditions.[15] But the prize for the most radical idea, in this regard, goes to Robert Freitas, an expert on nanotechnology at the Institute for Molecular Manufacturing in Palo Alto, California. The reason we eat food, Freitas observes, is so that our bodies can derive energy that plants and animals have preharvested for us from the sun. Why not bypass the plant/animal stage altogether, Freitas asks, and find a way to power our metabolism by directly plugging into a different energy source? The candidate he proposes is the radioactive isotope gadolinium-148.[16] If we could build nano-robots to course through our veins, assisting our body's cells with their basic metabolic processes, and if we could safely energize those millions of microscopic bots with gadolinium-148, then, in principle, such a system would suffice to power a human body through its daily activities. "Because gadolinium has a half-life of 75 years," Freitas notes, "humans might be able to go a century or longer without a square meal."[17] Scenario:

"Hey, you wanna go grab a bite to eat?"
"No, thanks. I just had a snack eleven years ago."

This proposal lies on the fringes of the plausibility spectrum—to put it mildly—but it does suggest the kinds of unorthodox possibilities that our society might choose to explore, as it seeks ways to sustain an enhanced human population over the coming century. The preparation and sharing of meals has always played a central role in human culture—a ritual of conviviality that has bound together individuals, families, and entire communities. As biotechnology is increasingly applied to food consumption, it holds the potential to unsettle those rituals and, perhaps, to dramatically destabilize them.

PRIVACY

In 2006 the writer David Holtzman conducted a little experiment: he went online to see how much information about himself he could gather from publicly accessible sources. Within a few days, he'd assembled a portrait that included his detailed physical characteristics (with photos), home address (with photos), educational background, jobs and employers, publications he'd written, family status, medical profile (including blood type and major diseases), police record, credit score, outstanding personal debt and mortgage loans, political contributions, favorite movies, books, and music, patterns of online behavior including daily e-mail load, personality type, and threat potential as measured by security agencies.[18] Some people, of course, *voluntarily* post a great deal of such personal information on social networking sites, but Holtzman's point was that this wide-ranging trove of data was available to one and all without his consent.

As more and more aspects of our lives move online, these seemingly isolated facets of our daily existence become, ipso facto, public or quasi-public information, captured by state-funded surveillance technologies and omnipresent informatic gadgets used by private citizens, generating immense databases, bound seamlessly together and rendered accessible by the Web. A few mouse clicks, and you can find the needle in the haystack in 0.0013 seconds.

This surely counts as one of history's major unintended consequences. Like the impact of World War I on the socioeconomic status of women, like the effect of the automobile on the social geography of cities, this outcome was never sought by any single person or group. It just, sort of, happened. We bought our various informatic gadgets, happily used them, and then discovered one day that the end result of all those gadgets being used was that we were losing our privacy. It was slipping away, irrevocably, just like that.[19]

Over the coming decades, the machines and devices with which we surround ourselves will become increasingly interactive and "smart," continually performing myriad small chores for us on their own automated initiative. Many such technologies already exist today in prototype form. Self-driving cars,[20] phones that dialogue with you, clothes with embedded microchips,[21] personable robots,[22] interactive homes connected to the Web,[23] toilets that analyze your stools[24]—all these devices either are for sale or are being honed and prepared for the consumer market. The informatics designer Adam Greenfield coined the term "Everyware" to describe this phenomenon. "What we're contemplating here," he observes,

> is the extension of information-sensing, -processing, and -networking capabilities to entire classes of things we've never before thought of as 'technology' . . . artifacts such as clothing, furniture, walls, and doorways. . . .

Designers are exploring how the possibilities inherent in an everyday object can be thoroughly transformed by the application of information technologies like Radio-Frequency Identification, Global Positioning System, and mesh networking.[25]

The proliferation of Everyware—referred to by some observers as the emerging "Internet of Things"—will do doubt bring all manner of new conveniences (and vexations) to our lives. What it will also do is dramatically increase both the quantity and quality of searchable data about you and your world. In the age of Everyware, there is no such thing as "off the grid" anymore. You are always a blip on the radar screen of some device or, more likely, of multiple devices that are all communicating with each other as they track and record your movements, body states, words, and deeds.

You can try to defeat or circumvent these technologies, of course. Give up your cell phone, drive your own darn car, refuse to wear "smart clothes." But the Everyware will outsmart you. Walk down the street, bearing no devices of any sort, and the various monitoring cameras at intersections, banks, or parking lots will still pick up your image: face-recognition software in the mainframes will be able to ID you and trace your movements. Insist on buying your own groceries, using good old untraceable cash dollars, and the checkout machine will still keep a record of what you bought, linking it to your customer file through the ID function of the cameras over the doorways, grocery aisles, and cash registers.[26] Take a walk in the woods, seeking to get away from it all, and the various interactive devices of other hikers will reach out and identify you, their networked connections faithfully recording your progress down the trails.

This all sounds, well, paranoid. But one doesn't have to be a black-helicopter conspiracy monger to see the overall trend here. As a society, we have embraced the many tangible benefits of ubiquitous, interactive, informatic technologies, and unless we are collectively willing to give up those benefits, we will simply have to accept the trade-off of dwindling privacy.

Some scholars have suggested constructive legal steps our society might take to mitigate this problem.[27] We should ensure that information can only be used for its stated purpose: if a sensor at airport security detects a tumor in my abdomen, this should not be allowed to affect my health insurance. When a software agent "decides" to put me in a certain category (such as a no-fly list or a low-credit category), I should be entitled to an explanation of why this has occurred and be given recourse to appeal the decision. There should be rigorous public oversight of any database that holds my personal information. All such databases should be subject to sunset rules: if the information they hold is no longer required for the purposes for which it was originally gathered, it should be permanently erased.

These are all sensible measures that should definitely be adopted. Nonetheless, they have a certain flavor of futility to them—like a child at the beach, frantically building walls and ditches around her sandcastle as she senses the tide rolling in. Such regulations will render our inevitable loss of privacy more tolerable, and that is perhaps the most that we can aspire to achieve.

THE ARTS

In 1849 the German composer Richard Wagner had a cool idea. In an essay titled "The Art-Work of the Future," he described something that he called the *Gesamtkunstwerk*, which translates roughly as "total artistic work."[28] The concept was straightforward: instead of writing poetry, performing music, carving a sculpture, putting on a play, or designing an architecturally exciting building, why not seek to do all these things at once, in a single work that harmoniously synthesized every form of creative expression? As the twenty-first century advances, the technologies of human enhancement will open up the possibility of creating artworks that are more mind-bendingly *gesamt* (all embracing) than even Wagner could have dreamed. Among the possibilities:

- New artistic media that use the technologies of virtual reality to create wholly immersive experiences going as far beyond today's cinema as a movie goes beyond a still photograph.
- Novel forms of sensation, like the magnetic dimension made possible by the feelSpace belt described in chapter 3, complementing and extending the four senses used in artistic works today.
- Drugs that alter people's perceptual and emotional reactions to external stimuli in carefully calibrated ways, allowing artists additional means for guiding their audiences toward new planes of aesthetic experience.
- Dream recordings (if feasible) that are interwoven into artworks, offering fresh perspectives on the interplay between conscious and subconscious levels of experience.[29]
- Innovative systems of audience participation and feedback, rendered feasible by advanced interactive features of bioelectronic implants or headsets, altering the traditional boundary between artist and audience.

- New kinds of collective authorship, akin to the remarkable cooperative creations one sees today in massively multiplayer online games.
- Brain-machine interfaces that allow artists to express themselves by directly transmitting and modulating subjective memories, emotions, impressions, or sensations—thereby offering a form of artistic communication that borders on psychoanalytic encounter or spiritual communion.

The advent of such exciting new forms of aesthetic expression would probably change not just how we experience art or communicate with each other, but would cut more deeply into our very sense of who we are—the boundaries of our individuality, the broader order of our social relations.[30] Unfortunately, this does not necessarily mean that all facets of mid-twenty-first-century culture will sparkle with deep innovation and creativity. In many cases, our advanced devices may accomplish little more than amplifying our greed and stupidity. In his novel *Feed*, for example, the writer M. T. Anderson envisions a future in which most humans are permanently connected to the Web via an implanted brain-machine interface—with 90 percent of the information flow consisting of cretinous advertisements that invade each individual's stream of consciousness.[31]

This latter scenario is, of course, a deliberately exaggerated one, but Anderson is not alone in sounding a cautionary note, for the remarkable technologies of the Internet and iPhone have already yielded a variety of troubling effects. Nicholas Carr, in *The Shallows*, argues that Web culture has done serious damage to our ability to concentrate and think deeply, substituting an incessant blur of rapid-fire trivialities for the quiet, sustained attention that people used to marshal in the days of print media.[32] Lawrence Lessig, in *Code: Version 2.0*, shows how the underlying architecture of the Web constitutes a framework of tacit "laws" that can constrain our freedoms in important ways.[33] Sherry Turkle, in *Alone Together*, highlights the paradoxical qualities of isolation and loss of meaning that have come to characterize human relationships in the era of ubiquitous informatics.[34] Jaron Lanier, in *You Are Not a Gadget*, makes a persuasive case that the Web has failed to live up to its original promise:

> Anonymous blog comments, vapid video pranks, and lightweight mashups may seem trivial and harmless, but as a whole, this widespread practice of fragmentary, impersonal communication has demeaned interpersonal

interaction. . . . A new generation has come of age with a reduced expecta-
tion of what a person can be, and of who each person might become.[35]

None of these critics of today's Web culture is necessarily arguing for a
Rousseau-esque return to some idyllic pre-plugged-in state of nature. What
they are in effect saying is, "Be careful, for these potent tools could end up
turning you into something you don't really want to be."

WARFARE AND TERRORISM

Over the coming decades, the ingenuity of human beings will no doubt continue to be applied, as it always has in the past, to devising ever more potent
ways to inflict harm on their fellow humans. We will probably see the advent
of nanotech weapons, epigenetic weapons, bioelectronic weapons, pharmaceutical weapons, AI weapons, space-based weapons, or robotic weapons.[36]
Some of these inventions may exceed the nuclear armaments of today in le-
thality. Many of them will be very expensive to make and difficult to conceal;
others, unfortunately, will be relatively cheap to make and easy to conceal.

The implications of this latter fact for national security are hard to over-
state. Nuclear weapons, for example, fall into the "hard to obtain and de-
ploy" category: they are made from rare materials, require large industrial
facilities for purification and enrichment, and leave a radioactive signature
that renders them relatively easy to locate. Other emerging categories of
advanced weaponry, however, are quite different. Highly contagious bioengineered microbes, for example, can be synthesized using materials that are
found in just about any high school or university biology lab. Handling them
does not require particularly sophisticated equipment or advanced knowledge, and they can be sealed away in small and undetectable vessels until
they are ready for unleashing.[37] This phenomenon of easy access will apply—alas—to a variety of increasingly powerful military technologies over
the coming century. It will have the unavoidable effect of "democratizing"
weapons of mass destruction, placing truly horrific devices into the hands
of ever-smaller groups of persons. A handful of lunatic extremists bearing a
grudge will be able to wreak unparalleled amounts of harm on tens of thousands, and perhaps millions, of innocents.

How our society deals with this problem will constitute one of the major challenges of the coming century or two. Citizens of democratic countries will face an unpalatable trade-off: either to grant their governments
unprecedented powers to monitor the private lives of their populations, in

the name of preempting terrorist attacks; or to hold on to their freedoms as much as possible, while accepting a certain level of vulnerability to fanatical minorities who are willing to kill multitudes to get what they want. This trade-off has never been easy to make in the past, but it will be even harder to handle when literally millions of lives are at stake.

What will the soldiers of tomorrow look like? Here we encounter the work of a federal research organization, the Defense Advanced Research Projects Agency (DARPA).[38] Created in 1958 as a response to the Soviets' launching of Sputnik, DARPA operates today with an annual budget of $3.2 billion (about half of 1 percent of total defense spending). It boasts an impressive track record, having helped fund the research that ultimately resulted in such inventions as stealth technology, e-mail, the Internet, the computer mouse, massively parallel computer processors, GPS systems, cell phones, weather satellites, fuel cells, laser weapons, and a variety of robots, ranging all the way up to the Predator unmanned warplane. Here is how the agency described itself in 2008 on its web page:

> DARPA's mission is to maintain the technological superiority of the U.S. military and prevent technological surprise from harming our national security by sponsoring revolutionary, high-payoff research bridging the gap between fundamental discoveries and their military use. . . . Since the very beginning, DARPA has been the place for people with ideas too crazy, too far out and too risky for most research organizations.[39]

Many of DARPA's projects focus on creating the superwarrior of tomorrow.[40] In this regard, the agency's assumptions are clear: human fighters running around with guns are already obsolete today. Future wars will be won by those military forces in which human beings have been most extensively and seamlessly interwoven with advanced machines—not just weaponry, but telecommunications, logistical support, sensing devices, robots, drones, and concealment technologies. The machine component in this pairing will make homo sapiens stronger, faster, tougher, stealthier, and more deadly than ever. The human component will render the machines more, well, sapiens.

This vision of the not-so-distant future has five main elements.[41]

1. *Physical abilities.* Soldiers will have modified blood in their veins, affording them considerably greater stamina and endurance, and the ability to operate effectively for long periods under extremely punishing

conditions. They will possess higher levels of dexterity and coordination, allowing them to move more nimbly and to handle more complex manual tasks than today. They will be able to operate for protracted periods without sleep, only negligibly compromising their alertness and effectiveness. When needed, they will put on flexible exoskeletons that allow them to trek long distances while carrying three-hundred-pound loads with minimal effort.[42]

They will be tougher and more resistant to wounds and illnesses. If they are wounded, they will possess the ability to survive longer without succumbing; once they reach medical facilities, they will heal more quickly and fully.

They will possess a wide variety of augmented senses, attuned to monitoring the environment for potential threats. Although it is hard to predict what those new portals of sensation will be, one can imagine, for example, the usefulness of infrared night vision, a telescopic zoom capability, image-sharpening circuitry, automatic brightness and glare reduction, heightened olfaction, receptors for detecting a variety of chemical traces, boosted tactile sensitivity, and the ability to tap directly into radio, magnetic, or electromagnetic fields. The soldiers of tomorrow will learn how to live in this enhanced "sensory sphere," constantly scanning the additional perceptual dimensions of their surroundings at a subliminal level. Their sense of selfhood, and of their place in the material world, may undergo a significant shift as a result.

2. *Cognitive abilities.* Soldiers will be able to stay focused for much longer periods than today, effectively resisting mental fatigue. A variety of pharmacological and bioelectronic interventions will increase their ability to construct mental models of their operational environment, analyze options, and make swift and effective decisions about how to act. They will possess a heightened spatial sense, allowing them to envision, mentally rotate, and analyze very complex situations and sequences of operations in their minds.

3. *Emotional control.* Soldiers will possess a greater resistance to the corrosive effects of stress, trauma, and fear than today: extremely dangerous conditions will not cloud their minds as strongly. Pharmaceuticals and bioelectronic stimulation will give them a heightened ability to compartmentalize their mental states, temporarily setting emotions aside in order to concentrate on the demands of the situation at hand.

4. *Interfacing with machines.* Soldiers will be equipped with cutting-edge bioelectronic devices, both invasive and noninvasive, that allow them to communicate seamlessly with machines, weapons systems, drones, and robots, controlling them by thought alone. Part of the training of military personnel will involve accustoming them to the constant informational demands required to control these machines in a fully transparent way: they will feel like natural extensions of the individual soldier's body.

Some specialized fighters will also possess bioelectronic devices that serve as weapons systems in their own right. For example, they may be able to jam the bioelectronics of their adversaries by projecting interference fields that block certain communications frequencies. One can also imagine more potent devices that allow an individual to "fry" an adversary's bioelectronic implants through a directed energy beam or electromagnetic pulse.

5. *Communications.* Soldiers will be able to maintain constant contact with each other, and with their officers, through bioelectronic implants or headsets that allow them to communicate swiftly and securely. These heightened real-time communications will allow them to inhabit a sphere of integrated information and situational awareness, giving them the ability to form far more integrated units of collective action than are achievable even by elite commando teams of the present day.

For those individuals who choose a military career, the phenomenon of professional specialization that I discussed in chapter 8 will take on particular significance. Part of the military's institutional mission is to fashion each individual warrior into the most effective killing machine possible. At the time of retirement, however, this creates a real problem. Many attributes of a killing machine are disastrously inappropriate in the context of civilian life, when the uniform is set aside and the soldier becomes once again a spouse, a parent, a colleague working a civilian job. The same enhancement technologies that will have rendered the soldier stupendously effective as an instrument of warfare will also make it all the harder for the retiring warrior to find her way back to her former, nonmilitary self.

. . .

Sex, food, privacy, the arts, and warfare—our brief excursion through these domains offers a concrete sense of the sheer *weirdness* that awaits us as we travel down the road of bioenhancement. These technologies open up exciting, luminous possibilities. Yet we would be foolish not to be at least a little bit scared as well. If you think your iPhone is a transformative device, just wait til they turn on your brain-machine interface.

Part IV

CHOICES

. . . . .

SIXTEEN

Why Extreme Modifications Should Be Postponed

Or, the Singularity Can Wait

Thou art the molder and maker of thyself; thou mayest sculpt thyself into whatever shape thou dost prefer.

—God to Adam, in Pico della Mirandola, *On the Dignity of Man* (1486)[1]

STEERING THE EVOLUTION OF *HOMO SAPIENS*

Suppose you are a seventeen-year-old making plans for college. You know you are making decisions that will shape the trajectory of the rest of your life, yet you do not let the far-reaching implications of your choices paralyze you. You choose which colleges to apply to, which parts of the country you'll be living in, what kinds of major you'll probably be considering. In making these plans, you realize full well that your values, goals, and preferences will probably evolve somewhat over the decades that follow. The journey itself will transform you, and the destination will undoubtedly change many times over. But you nonetheless have to make your college decisions today, one way or the other, and require concrete principles to guide you right now. You therefore make a provisional judgment, going forward with your current best estimate of what kind of person you are, and what kind of person you want to make of yourself over the years to come.

Our species finds itself today in an analogous position. The advent of bioenhancement technologies means that the contours of human identity will probably shift significantly over the coming century. Whether we are ready or not, we all face a similar set of future-oriented choices: What sort of creature do we want to make of ourselves? And equally importantly: What sort of creature do we want to avoid becoming?[2]

As our society embarks on this transformative journey, we need clear ethical guidelines—however provisional in nature—to lend coherence and meaning to these far-reaching decisions.[3] The best candidate for such an ethical framework, I have argued, is the cluster of ten concepts described in chapter 6, under the heading, "Ten Key Factors in Human Flourishing": security, dignity, autonomy, personal fulfillment, authenticity, the pursuit of practical wisdom, fairness, interpersonal connectedness, civic engagement, and transcendence.[4] These values can guide us as we grapple with the core long-term choices posed by enhancement technologies: How far do we want to go in altering the human constitution? What kinds of modifications should we embrace, and which should we reject?

I turn now to these "species-shaping" questions, assessing the space for human agency in determining our own biological and cultural future.

WHAT IF WE JUST SAY NO?

Let us suppose, for the sake of argument, that a majority of US citizens were to decide today that human bioenhancement is a bad road and that our society should refuse to go down it. Could we stop these technologies from being developed? What would such a policy of across-the-board relinquishment require?

The answer is sobering. Since a significant minority of citizens would presumably wish to press ahead with self-modification, stringent new legislation would be required to prevent them from doing so. One would then expect to see the emergence of a widespread black market for enhancement devices and procedures, harkening back to the era of Prohibition in the 1920s. In addition, banning all major forms of enhancement would require selectively blocking key areas of research in many fields we value highly, such as computers and informatics, genetics, robotics, neuroscience, nanotechnology, cognitive psychology, pharmaceuticals, and cutting-edge medicine. The surveillance apparatus required to enforce the ban would have to approach totalitarian levels in its fine-grained comprehensiveness. Finally, of course, a successful ban would require this highly restrictive system to be imposed equally scrupulously in all the world's nations at the same time, or you would find the research and innovation simply migrating overseas to the least-regulated regions of the planet.[5]

For all these reasons, therefore, the idea of enacting an across-the-board ban on enhancement technologies is a nonstarter.[6] But this hypothesis of a full-scale ban is itself rather far-fetched, because the much likelier prospect is that a great many citizens will eagerly *seek out* such modifications, and (once proven safe and reliable) purchase and employ them with gusto. We can expect a broad spectrum of opinions about enhancements to animate the citizenry of the coming century—from total rejection at one extreme to enthusiastic embrace at the other. Some forms of enhancement (like disease resistance) will no doubt appeal to all but the most die-hard purists; other forms (like brain implants) may only prove desirable to certain subgroups among the population. Therefore, legislators and policymakers will need to determine which kinds of modifications to allow and which ones (if any) to prohibit. They will also have to find ways to regulate effectively those forms of modification that do come into widespread use.

HOW FAR TO GO?

One of the tough decisions that our society will face is whether to stick with relatively modest and restrained forms of self-modification or to proceed unabashedly into more extreme territory. If we humans acquire the ability not just to tweak our own biology, but to redesign it entirely, should we do so? Are there certain lines and thresholds awaiting us out there that we should be wary of crossing?

Implicit here lies a distinction among three levels of possible human enhancement.[7]

- *Low-level modifications: Capabilities at the high end of today's human range.* If you play tennis as well as I do (and the only sporty thing about me is my athlete's foot), low-level enhancements would result in your playing like Roger Federer. If your IQ is 110 today, you would end up thinking like an Einstein or a Picasso. Your life span would be boosted to about 100 or 110 years of age, but you would still show all the telltale signs of the aging process.
- *Mid-level modifications: Capabilities well beyond today's human range, but still recognizably human.* You would play tennis vastly better than Federer— more like the fictional characters described in my first vignette, with catlike reflexes, superhuman strength, and so on. Your cognitive skills would take you beyond even Einstein and Picasso, into new realms hard to imagine by today's minds; nevertheless, your ideas would still make sense, after a great deal of patient explanation, to today's physicists and artists. Your health span would extend to 150 years or more; rejuvenation treatments would keep you looking and feeling like a forty-year-old, despite your much greater chronological age.

Notwithstanding the unprecedented nature of these attributes, no one could legitimately question your underlying humanity, because the overall profile of your capabilities would still remain recognizably commensurate with the traditional human range of behaviors, experiences, and aspirations.

- *High-level modifications: Capabilities utterly beyond human parameters.* You would not wish to play tennis at all, because no human player (enhanced or not) could hold a candle to you. The game would be as boring for you as it would be for one of today's humans to repeatedly staple sheets of paper together for three hours in a row. Your mind would be so advanced that your ideas would be impossible for even the smartest humans to grasp: your thoughts would entail levels of dimensional dexterity, mathematical nuance, immense working memory, and sheer conceptual complexity that would place them hopelessly beyond even the *possibility* of human comprehension.

This latter form of high-level metamorphosis appears to be what many transhumanists eagerly envision for themselves. Thus, the website of the transhumanist organization Humanity+ posits: "Some posthumans may find it advantageous to jettison their bodies altogether and live as information patterns on vast super-fast computer networks. Their minds may be not only more powerful than ours but may also employ different cognitive architectures."[8]

Now imagine: the year is 2099, and you are ready to transmogrify yourself. You step into the "H+ apparatus," and the technicians outside power up the controls. The procedure begins; you feel your old identity rapidly melting away. Your powers grow, your mind expands, your emotions open outward. Let us suppose, for the sake of argument, that the procedure endows you with a massively parallel cognitive architecture that allows you to handle disparate streams of information simultaneously. Thus, you now find yourself able to undertake all ten of the following activities at the same time:

- Read Conrad's *Heart of Darkness* streaming from the Web.
- Balance your checkbook.
- Practice yoga.
- Have a phone conversation with a friend.
- Work on a mathematical proof.

- Design the future layout of flowers and shrubs in your garden.
- Create a recipe for pumpkin pie with maple topping.
- Revise a poem you wrote yesterday about raindrops.
- Practice your Mandarin Chinese.
- Fantasize about engaging your lover in a long, sensuous kiss.

It is worth underscoring: you are *not* switching rapidly to and fro, paying attention to each of these activities separately, then moving swiftly over to another one. Rather, you are wholly engaged in each of them while doing *all of them* in truly parallel, simultaneous fashion.

What do the concepts of "here" and "now" mean to such a mind? These are two concepts that characterize ordinary human consciousness, which is always necessarily limited to the focalized handling of one object of attention at a time: "I am here now, sitting at my computer, thinking about this specific problem." By contrast, such a parallel-structured mind would conceive of, and experience, its own awareness in profoundly different ways from today's humans. It would not exist in a here and now, but would exist simultaneously, like the Beatles song, "Here, There, and Everywhere."

Would such a form of distributed consciousness be compatible with human personhood?[9] Probably not. It would have no proper center and, therefore, no unitary self. It would in fact have multiple simultaneous sub-centers, each perfectly capable of going its own way, having its own experiences, engaging in its own interactions with the world around it. Yet one of the essential features of human consciousness lies precisely in its possession of a unique reference point—represented by the word "I" in every human language—in connection with which all one's experiences and deeds stand related.[10] Despite the many differences between the kinds of knowledge produced by neuroscientists and philosophers, they all converge on this basic principle of focalized selfhood.[11] A being whose mind could be wholly immersed in many disparate activities at the same time would possess a de-centered, dispersed form of awareness that would, by definition, preclude such focalized, unitary selfhood from existing.

Human personhood, in other words, arises out of a specific configuration of capabilities *and limitations*. Our incapacities, blind spots, weaknesses, and functional constraints define our being just as significantly as our positive powers and faculties. The inability to exist in many places at once is just as important to our identity as our ability to think logically or to experience emotions.[12]

What the transhumanists are saying, in effect, is: "How nice it would be to still be me—but to have a radically expanded cognitive architecture."[13] They apparently do not grasp the fact that, in order to acquire those awesome capabilities, one would have to alter one's body and mind in profound ways, and those alterations would in turn transform one's identity, so that the resultant successor entity would no longer remotely be oneself. The scenario involves an unavoidable trade-off: you can have the radically augmented capabilities *or* a recognizably human form of personhood, but not both. What the transhumanists are really saying, in other words, is: "How nice it would be to terminate my selfhood, bringing into existence an utterly new kind of sentient entity."[14]

POSTPONING EXTREME FORMS OF TRANSMOGRIFICATION

A libertarian might argue that the project of undertaking extreme self-modification is like that of an individual undertaking assisted suicide: society may disapprove, but it should not meddle in the decisions that a private citizen makes regarding his or her own continued existence. If I want to die and have reached this decision after long and careful deliberation using my full mental faculties, I should not be prevented from bringing this about.[15] We cannot be sure whether extreme transmogrification would cause my biological death, but if this is a risk I undertake willingly, then society has no right to stand in my way—as long as my radical transformation occurs in a manner that harms no one else.

Ah, but there's the rub. Unlike the assisted suicide of a single individual, which arguably poses no direct danger to the rest of society,[16] the act of undergoing extreme transmogrification inevitably entails serious risks, not just for the person doing it, but for the rest of humankind as well. Such acts of creation would bring into being new kinds of "posthuman" entities that have the potential of being extremely powerful and uncontrollable. We have no way of knowing how they would behave toward the rest of the biosphere—including all other sentient beings on our planet. We do have good reason to believe, however, that they would possess capabilities far beyond those of today's humans and that, as a group, their powers would far surpass those of humankind.

Since their minds will be so profoundly different from ours, we cannot be sure that they will share our ethical standards for proper treatment of fellow creatures and other species. They may cherish us old-fashioned humans and treat us with great benevolence. They may be oblivious toward us, just as we are toward the microbes in our intestinal system. They may regard our bodies as convenient sources of trace elements to be mined like a vein of ore in a hillside. The plain fact is that we will have no way of knowing what they are like until we have actually created them and interacted with them. By then, of course, if things go badly, it will be much too late.

This seems to me like a no-brainer. Blindly rolling the dice with the future of all life forms on our planet is not a good idea. Until we know a great deal more than we do today about what such entities would be like, and how we might be able to assure a state of peaceful coexistence with them, it would be the height of folly and irresponsibility to proceed with the project of creating them. Above and beyond all the moral questions, this matter boils down to a simple prudential argument. The potential rewards are too uncertain, and the risks far too great.[17] A day may come when we feel confident that we sufficiently understand the underlying dangers and benefits of creating such beings—a day when our citizenry achieves a reflective and democratically reached consensus that it is wise to proceed. But until that day arrives, we should defer the Singularity.[18]

AVERTING FRAGMENTATION OF THE SPECIES

Apart from these basic safety concerns, moreover, there is a second good reason for postponing extreme modifications: their cultural consequences could end up tearing our society apart. Let us suppose that, a few decades hence, millions of people throughout the planet have begun exploring the thrills and challenges of the enhancement enterprise. One possible consequence, which I explored in chapter 8, is that these bioenhanced humans would spontaneously sort themselves over time into discrete social clusters based on particular trait ensembles and performance profiles. Such groupings might start out as informal gatherings of loosely connected persons who have noticed a basic affinity in their chosen trait profiles. But with the passing of decades, it is possible that they will form increasingly self-conscious and explicit categories, bound together by common interests and values, as well as by a shared culture and ideology.

If this kind of clustering effect begins to manifest itself among the population of the enhanced, it will be vitally important for citizens to continue to recognize each other's common humanity. Even though you belong to a separate cluster from mine, and your bioenhanced capabilities differ from mine in significant ways, I need to see you still as a fellow human—someone whose personhood, dignity, and fundamental worth remain beyond question. *But if our society proceeds too quickly in allowing radical and outlandish modifications to be undertaken, this sense of underlying commonality would be much harder to sustain.* Persons in other clusters might start to appear not just as different but as downright alien. The result could be a new and extreme form of "identity politics"—a global society increasingly riven by negative stereotypes, resentments, rivalries, and outright conflict.

Here, then, are two excellent reasons for proceeding slowly with bioenhancements: the creation of powerful posthuman entities risks straining our social order to the breaking point, and it may endanger our very survival. But will these two reasons be sufficient to keep humankind from plunging headlong down this path anyway?

The answer is far from clear. Deep impulses will undoubtedly come into play, at both the personal and societal levels, pulling many among us toward ever more intensive forms of self-modification. Some of these propelling factors will reflect our baser nature: greed, competition, envy, and the lust for power. Others will arise out of noble sentiments: the desire to see our loved ones succeed; the thirst for novelty; the aspiration to attain higher forms of achievement, knowledge, and sensation. These forces will be hard enough in themselves to resist, but they will be further strengthened by the heavy involvement of large-scale business interests, for whom the "enhancement industry" will offer major profits. Influential libertarian voices will also add to the mix, as they invoke the inalienable right of each individual freely to modify her own body and mind as she sees fit. This nexus of impulses and ideals, economic and social forces, will generate a seemingly irresistible pressure to go faster, faster, faster.

And yet, I have argued, restraint is the smarter path: the deliberate postponement of radical forms of self-modification until our society has had a chance to gauge the consequences and acclimate to them. Unfortunately, as the historical record abundantly shows, just because an option is demonstrably smarter, this does not mean that people will choose it. They often invest recklessly, fail to plan ahead, pollute their habitat, or exhaust their resources: the dumb path is a well-traveled road.

Let us assume, for the sake of argument, that a majority coalition of citizens and associations in our society comes to be persuaded eventually by the arguments for caution and restraint. What concrete policies might they seek to put into place? One option would be a phased moratorium on radical bioenhancements—a middle road between banning these technologies and allowing them to proceed unregulated and unfettered. This would be in many respects a difficult policy to enact, but in its broad outlines, it might offer a way forward.

Our society might start by passing laws that allow only low-level enhancements to be undertaken—those that yield capabilities at the high end of today's human range, along the lines of Federer and Einstein. These legal restrictions on trait modifications would be framed as provisional and incremental in nature, not as absolute and permanent. After a certain interval of time has passed—perhaps several decades, perhaps considerably longer—citizens and legislators could revisit the issue: Have our socioeconomic and cultural systems adapted well to the challenges posed by these new generations of modestly enhanced humans? To the extent that a clustering effect has manifested itself, has it remained within manageable limits?

If the collective answer to these kinds of questions tends toward a clear yes, then the restrictions on enhancement technologies could be incrementally relaxed. Certain kinds of mid-level modifications might be gradually legalized, allowing people to give themselves new capabilities beyond today's human range.[19] Radical transmogrification would continue to remain off-limits, but humans would begin to probe new heights of performance like that of the fictional tennis player Felipe Cardozo in my earlier vignette. Certain forms of advanced brain-to-brain communications media (if feasible) might enter the market. Cognitive enhancements would push beyond the levels associated with the great geniuses of the past, exploring new conceptual dimensions and landscapes.

Admittedly, enforcing such a phased moratorium would pose significant legal and practical challenges. A sizable minority of the population might still be keenly interested in experimenting with radical self-modification, and one would therefore expect a black market to develop for such powerful fringe technologies. The police and criminal justice system would thus be presented with yet another arena—like today's drug wars—of costly ongoing struggles to rein in certain dangerous human activities. Legal scholars would no doubt have a hard time defining the line between allowable and illicit modifications, for in many cases, that line will be a blurry or movable one.[20] Many scientists and medical researchers would have to accept new levels of surveillance of their laboratories, imposed by the government as it seeks to prevent hazardous modifications from being carried out in secret.[21]

These difficulties are daunting enough, but orchestrating an effective international moratorium on all but low-level enhancements would pose an even greater challenge. A strong coalition of leading nations would have to emerge, all reaching consensus that such a moratorium was vitally important to their future. In order for such a consensus to become a reality, the citizens and leaders of those countries would need to agree that an unregulated international arms race in robotics, AI, and radical bioenhancements was against their long-term national interest. Such a consensus would be hard to reach, but it is not without precedent, as the history of arms control negotiations and treaties can attest.[22] If such a coalition were to form, its members could presumably exert all kinds of mounting diplomatic and economic pressure on other nations to follow suit. As more and more countries came onboard, any refuseniks would become pariah nations and would have to pay an increasingly high price in sanctions and isolation if they stubbornly refused to join the others.

But what if a rogue nation like North Korea were to defy the international consensus and proceed with the development of superintelligent robots or transmogrified citizen-soldiers? In such a case, a preemptive war, followed by invasion and forcible regime change, might not be out of the question. If other nations came to perceive the threat from noncompliance as sufficiently significant, they might conclude that this kind of drastic enforcement action was justifiable. Needless to say, a preemptive war of this sort would have to take place very early in the technological development cycle, well before the superrobots or superwarriors conferred a decisive military advantage on the noncompliant nation.[23]

One factor in favor of such an international governance system would be the phenomenon of waning privacy I described in chapter 15. As the decades of the twenty-first century go by, our society will be permeated by so many interactive, semi-intelligent informatic devices that it will be hard for anyone's actions, words, or endeavors to remain inaccessible to outsiders. This is a major bummer, to be sure. But the flip side is that many forms of unlawful activity will also become increasingly difficult to conceal. If an unscrupulous biologist decides to open up shop in the basement of his home, setting up a lucrative black-market business in illegal enhancements, it would become hard for such an individual to operate under cover of secrecy for long. Even if he managed to keep his advanced technologies under the radar, many of his clients themselves would eventually be found out and prosecuted. Police would be able to trace their past activities and movements, and soon ferret out the source of the illegal modifications. The high likelihood of being caught would constitute a strong disincentive to anyone who wished to thwart the aims of the moratorium.

Still, it is conceivable that some clever individuals or groups may succeed in evading this ubiquitous informatic net and proceed with their plans to create a superintelligent AI or a powerful form of posthuman being. Law enforcement agencies at all levels—municipal, national, and international—would therefore need to have contingency plans in place for responding to such a rogue development with swift and implacable force. This would add yet another element of basic uncertainty to the already impressive set of threats looming over mid-twenty-first-century society: the possibility of an unstoppable new technological or biotechnological creation coming out of nowhere and dictating its terms.

All these difficulties, obstacles, and drawbacks suggest that the smart path of restraint would be an arduous one to follow, even if a majority of citizens endorsed it. It would be messy, contentious, frustrating, and fraught with all manner of intractable practical and legal challenges. But what is the alternative? Muddling through, full speed ahead, without any guidelines at all? Keeping regulations to a minimum, letting the market decide, and just waiting to see what happens? The risks involved in these technologies are potentially catastrophic in nature: what is at stake is the integrity of our social order, and possibly our survival as a species. We simply cannot afford a hands-off approach, when the potential dangers are this profound. We will have to do our best to confront each of these difficulties and obstacles, one by one, gradually piecing together a policy of restraint that brings some elements of structure and control into this process as it unfolds.

SEVENTEEN

Humane Values in a World of Moderate Enhancements

I've gone just about as far as I can go with this body.

—Linda Litzke (Frances McDormand)
in *Burn After Reading* (2008)[1]

We now turn to the relatively more modest forms of bioenhancement that our children and grandchildren may find themselves experiencing over the decades to come. I am referring here to the low-level and mid-level modifications described in the last chapter: those that yield capabilities at the high end of today's human range (Federer, Einstein) and those that push a bit further, into territory beyond today's human range (but still recognizably human in nature). No one can know, of course, what specific forms of bioenhancement will be adopted by the coming generations, but it is worth playing out some of these potential scenarios in a speculative mode, laying out the kinds of problems that seem most likely to arise and the strategies that may be required for mitigating them.

The benefits of these coming biotechnologies are likely to be far-reaching. Physical and mental rejuvenation, greater control over our bodies and emotions, improved cognitive abilities, novel modes of communication, different ways of interacting with machines—these are only a few of the possibilities coming over the horizon.[2] At the same time, as we have also seen, the challenges and drawbacks generated by these new capabilities will no doubt be considerable as well: we risk undermining some of the qualities we value most highly in our social order and in our spirit as individual persons. The question then becomes: Can our society find ways to reap the benefits while avoiding or diminishing the dangers?

MITIGATING THE DOWNSIDES: SEVEN KEY CHALLENGES

1. Radical inequality

If the most effective enhancement technologies are prohibitively expensive and are therefore accessible only to society's elites, a new form of insurmountable inequality would be likely to emerge. Some people would be able to prodigiously outperform other humans, running circles around them in basic health, physical performance, mental acuity, communication, and ability to interact with advanced machines. In chapter 7, I described this outcome as a sort of caste system inscribed into biology itself—a hierarchy based not on socioeconomic achievement per se, but on the brute fact of overwhelmingly superior capabilities. If this form of "biostratification" were to occur, the principle of equality of opportunity—a key feature of any genuine democracy—would become permanently unworkable. Since such a dire outcome would probably prove unacceptable to a majority of the citizenry, I argued, our society will have no choice but to implement a system of universal access, offering subsidized availability of a "basic package" of enhancement technologies to all citizens. Under such a system, no one would be compelled to adopt enhancements, but every citizen would have the option of incorporating the full range of legal enhancements if she wanted them.

National governments will probably have to play a significant role in establishing the system for universal access. Either they will run the distribution of enhancement technologies directly, or they will regulate (and partially fund) the distribution mechanisms operating in a free-market context. Four basic challenges immediately present themselves in this regard.

HOW WILL GOVERNMENTS PAY FOR IT? If these technologies are very expensive—and it is likely that some of them will be—how can they be made available to all, without causing huge deficits in government balance sheets? A good place to seek an answer is in northwestern Europe, among the social-democratic welfare states such as Sweden, Denmark, or Germany. It is all a question of priorities. Comprehensive education and health care are very expensive as well, but those nations have found ways to strike a balance—providing a broad array of high-quality basic services to all, while still maintaining responsible levels of expenditure year after year.[3] In all likelihood, a similar type of balance will need to be developed for the enhancement technologies. If citizens feel strongly enough about avoiding the emergence of a caste system in their society, then they will have to make room in their national budgets for the financing of universal access. This may require a recalibration of traditional spending priorities—cutting back for example on social entitlements, public projects, or military outlays. On the other side of the ledger, of course, it is worth noting that many forms of bioenhancement can be expected to greatly increase the overall health and productivity of the citizenry, thereby expanding the economic resource base on which future societies can draw for funding universal access.

AVOIDING EUGENICS. The policy of subsidized access will no doubt apply only to a specific array of mainstream enhancements. If individuals wish to install more outlandish or esoteric modifications, they will probably be required to pay for these on their own. Certain stringent guidelines will also have to be established, to ensure that the traits and capabilities adopted by people pose no significant danger to others. For example, a weaponized bioelectronic implant, capable of frying other people's implants by beaming a strong electromagnetic pulse at them, would fall under this category. Although military personnel might make use of such devices, they should not (*pace* the National Rifle Association) be allowed for civilians.

Since the central government, in this scenario, takes on a key role as either a provider or regulator of enhancements, it will unavoidably be in a position to channel the choices made by citizens, incentivizing some types of modifications while discouraging others. This would constitute, ipso facto, a form of state-sponsored eugenics, reminiscent of the policies adopted by a variety of Western governments during the first half of the twentieth century.[4] Those early eugenics programs have come to be recognized in subsequent decades as egregious violations of human dignity, so it will be important to establish a system of checks and balances, with rigorous oversight, so that the role of government remains as neutral as possible in this regard. The goal, for government regulators, should be to maximize the range of safe options open to individuals and families, with heavy emphasis placed on the freedom to choose.[5] A powerful and independent oversight agency, working closely with citizens' groups and other nongovernmental organizations, would also be needed to monitor the government's ongoing policies and regulatory decisions. Only in this way can society avoid repeating the moral tragedies of the early eugenics movement.

GLOBAL DISPARITIES. It will prove hard enough, in the *rich* countries, to find resources for financing universal access; for the poor ones, the outlook appears grim indeed. Many Third World nations are unable to provide their citizens with basic nutritional needs, health care, and education. The idea of offering advanced drugs, bioelectronics, and genetic technologies to people living at such bare subsistence levels seems nothing short of absurd.[6]

I see two kinds of possible outcomes here. The nightmare scenario is a biologically bifurcated humankind: the enhanced citizens of the affluent nations, rising steadily to ever higher levels of capability and prosperity; and the unenhanced (or weakly enhanced) in the Third World, confined to the sidelines of history, mired in relative stagnation and despair. On the other hand, a more hopeful possibility also presents itself. Might it be that the sheer awfulness of the nightmare scenario finally awakens a sufficient level of motivation among the world's nations to tackle the problem of global poverty in an effective manner?

An intriguing precedent, in this regard, is the Montreal Protocol of 1987.[7] This treaty, ratified by 197 nations, stands as one of the great success stories of international cooperation and solidarity: it has effectively addressed the problem of ozone holes created in the earth's atmosphere by the release of chlorofluorocarbons (CFCs) in certain types of aerosols and refrigeration systems. The rich nations not only phased out the use of CFCs within their own borders, but also set up a special fund to assist the poor nations in doing the same thing as well. Britain's Margaret Thatcher (not known for softheartedness toward the poor) took the lead in pushing the treaty through: she recognized that the danger was great, and that the Third World nations simply lacked the wherewithal to convert their economies away from CFCs without outside help. Faced with a serious global threat, in short, the rich nations ponied up to assist the poor nations in addressing the common challenge.[8]

It is conceivable that the threat of the nightmare scenario described above might suffice to motivate people in a manner similar to the threat of ozone depletion in the 1980s. If that were to happen, then perhaps the rich countries might join together in creating a new, global version of the Marshall Plan, designed to lift the Third World decisively out of poverty.[9] Such an intervention would require—like the original Marshall Plan—an extraordinary level of enlightened self-interest among the citizenry of the affluent world.[10] Perhaps this is only a fool's hope—but then people said the same thing to Marshall in 1947 when he first proposed his idea.[11] Whichever way it goes, one conclusion seems clear: unequal access to enhancement technologies, on a planetary scale, would pose a serious threat to the continued unity of the human species.

GOVERNING THE RELATIONSHIP BETWEEN THE ENHANCED AND THE NON-MODS. Even with a system for universal access in place, many individuals and families will steadfastly refuse all major forms of modification. I suspect that these non-mods will constitute a relatively small minority of the population, especially after many decades go by and new generations are born for whom enhancement technologies are an established, "normal" feature of the world in which they grow up. Nevertheless, our society should recognize a person's decision to refuse bioenhancement as a basic human right. Strong government policies will be needed, both to protect such non-mods from discrimination at the hands of the majority, and to make sure that they are afforded the basic socioeconomic conditions required to provide for themselves and their families.

2. Species fragmentation

Even if a phased moratorium on radical enhancements were adopted (see chapter 16), the coming generations may still find themselves facing a significant clustering effect among different subgroups of the moderately enhanced population. In order to avert such a potentially toxic outcome—or at least dial it down to manageable levels—our society would need to adopt a variety of "counter-clustering" strategies. Here are two ideas that might help.

INSTITUTIONALIZE CROSS-CLUSTER COLLABORATION IN GOVERNANCE. History provides numerous examples of polities in which multiple ethnicities and cultural groupings have lived and worked side-by-side for long periods in relative peace.[12] One thinks of the Habsburg monarchy in the nineteenth century, or nations like Belgium and Switzerland, or city-states like Singapore. In all these cases, the potential for strong centrifugal forces existed, yet those forces were counterbalanced by wise policies that actively bound the various constituencies together. No single group was allowed to predominate, or even to be perceivable as predominating. Equality before the law was vigorously enforced. Functions of day-to-day governance were subjected to power-sharing mechanisms that included all major stakeholders.

Of course, one can just as easily summon up plenty of historical cases in which this kind of multicultural order ultimately fell apart—often culminating in a bloodbath. Yugoslavia and Rwanda are only the most recent specimens.[13] But if it turns out that enhancement technologies eventually do bring about the clustering effect described above, then it seems a pity not to learn from history's success stories of interethnic integration. Wise leadership and well-crafted institutions can go a long way toward mitigating the centrifugal tendencies of a biologically divided polity.

AGGRESSIVELY PROMOTE AN ETHOS OF TOLERANCE AND RESPECT ACROSS CLUSTERS. Precisely because the morphological diversity brought on by enhancements would cut even more deeply than the ethnic divisions of today, our society will need to adopt a particularly vigorous stance in educating its citizens to reject all forms of prejudice and discrimination.[14] Novels like *Huckleberry Finn* have taught generations of black and white American children about elemental human dignity; programs on the Holocaust and other genocides have shown the post-1945 generations what happens when prejudice is allowed to run amok.[15] The school curricula of the coming century will need to further deepen this ongoing endeavor, adapting it as necessary for the radical new divisions that enhancement technologies may be expected to generate.[16]

A powerful tool, in this regard, may come from the bioenhancements themselves. As we saw in chapter 13, affordable virtual reality devices may be able to deliver extremely realistic simulated experiences in the near future. Such a technology might allow individuals to place themselves temporarily in the shoes of a person from a separate enhancement cluster, experiencing the world from that person's perspective. I could momentarily shed my own particular suite of modifications and try on yours for size: What's it like to be augmented in such drastically different ways? Carried out systematically, and accompanied by critical discussion and reflection, this kind of wide-ranging exposure to the lifeworlds of other people could bring about a significant increase in an individual's capacity for empathy. For some persons, at least, it might help build bridges of shared feeling and mutual understanding from one cluster to another.

3. New kinds of suffering

In chapter 9 I described various kinds of torment that might result from our tampering with the physical and mental constitution of animals. This included unprecedented forms of anguish that could be experienced by mammals or other sophisticated creatures whose cognition has been boosted to humanlike levels. Our society should place a high priority on preventing such abuses from happening. A first step would be to create a governmental ethics board empowered to establish clear guidelines for the humane modification of animals; those guidelines would form the basis for new legislation that would be binding not only on research laboratories but on commercial enterprises as well. A special agency would need to oversee the implementation and enforcement of these laws; and the participation and oversight of nongovernmental organizations should also be encouraged.

Suppose, for example, that a biotech company were to create several hundred "augmented orangutans" like Jeremy in my earlier depiction. Such beings would probably embody a new kind of personhood, and we would be morally obligated to accord them significant rights, commensurate with their minds, capabilities, and emotional profiles. I doubt that human society is remotely ready at present to make room for this novel category of citizen in our midst—the animal-person. We would therefore be well advised to postpone such forms of uplift until we humans ourselves are psychologically and philosophically prepared to share the governance of our polity with our hybrid creations.[17]

4. Defending mental privacy

It is possible that scientists will never succeed—for purely technical reasons—in creating a device that allows us to read another individual's thoughts without that person's consent. Still, the rapid advance in recent years of neuroscience, cognitive psychology, and biomedical engineering gives plenty of grounds for concern. Mind-scanning technologies are already starting to appear in relatively crude form today. They are growing in potency and accuracy with every passing year, and their trajectory points toward ever-increasing capabilities for monitoring the contents of a person's mental activity. The technologies of covert or forcible mind reading may not be a sure thing, but we have ample reasons for taking the possibility seriously.

What would our lives be like without the guarantee of basic mental privacy? A key aspect of our individual selfhood rests on the distinction between our inner world of subjective experience and the external reality in which we interact with physical objects and other people.[18] Forcible or covert mind-reading devices would severely degrade this distinction. They would take our innermost nuances of feeling or deliberation—our plans, desires, memories, intimate thoughts, spontaneous impressions—and place them on the table for all to see. Moment by moment, you would become vulnerable to giving away your intentions or judgments in ways that could lead to harsh consequences. You would therefore need to become far more vigilant and guarded in "managing" the flow of your own thought process: *"Here comes my boss: I'd better find a way not to think about the affair I know he's been having."* Such a technology would (literally) turn our familiar world inside out, destroying the foundation of self-contained interiority that is the crucible of our very personhood. We simply cannot allow this to be taken from us.

A capability for forcible or covert mind reading would obviously constitute a revolutionary instrument in the hands of government agencies like the FBI or CIA. But even if using such a device allowed our security agencies on occasion to thwart terrorist attacks and save a significant number of lives, our society should still draw a clear line and say an emphatic no. What good is saving specific persons' lives, when the technology that allows us to do so would grievously undermine the quality of *all* persons' lives, everywhere? The right to mental privacy is an absolutely valuable quality that should never be placed in any cost-benefit trade-off. This is a prime example of a form of technology that might seem to offer certain momentary advantages, but that would on balance bring overwhelmingly evil consequences to our world. No one should ever be allowed to possess such a device—no loopholes, no exceptions.[19]

5. Commodification

"Can't buy me love," runs the Beatles song. The Fab Four clearly grasped a basic truth that some economic theorists unfortunately do not: some things should not be put up for sale.[20] Two recent books by the scholars Michael Sandel and Debra Satz make the point very well.[21]

- *You can buy a wedding cake, but if you are the best man, you should not hire someone to write your toast to the groom and his new wife.* That toast is supposed to convey the affection that you feel, arising from your own long friendship with the groom, not the generic sentiments that an outsider would express on your behalf. The fact that you have to struggle to articulate those feelings, laboring to find the right words—however imperfect—are part of the gesture's meaning.[22]
- *You can buy a book for your kid, but even if she is reluctant to take it up because she's spending night and day online, you should not pay her twenty bucks to read it.* You want her to learn how to experience the pleasure of reading for

its own sake—not to slog her way through books on a cash-per-page incentive.[23]

■ *Donald Trump has a lot of money, but if he tried to buy the Statue of Liberty in order to place a flaming "Trump" sign on the torch, most Americans would balk.* Some things belong to all of us, as a community, and their symbolic resonance would be vitiated if they were appropriated for a single group's purposes.[24]

These examples illustrate what happens when people confuse two very different registers of value: instrumental usefulness as opposed to intrinsic meaning.[25] A pizza's principal purpose is to sate your hunger and give you pleasure: because its value is instrumental, there is nothing wrong with paying somebody to cook one for you. But in the three examples above, we encounter phenomena that matter to us for a very different set of reasons. Friendship, the pleasure of reading, the affection we feel for a communal symbol dedicated to the idea of freedom—these are aspects of our lives that arise spontaneously, infusing our experience with rich significance. They fulfill no narrow function, but partake of the ineffable realm of qualities that make life worth living. These types of experiences can never be subject to a calculus of quantifiable value, for they are unique, intimately meaningful, and freely arising gifts.

Money, when applied to such intrinsically valuable entities, not only fails to achieve its purpose; it undercuts their validity and robs them of their power to move us. It cheapens them, by implying that they can be compared or measured, bought or sold like any common object. Herein lies the essence of commodification.[26] Whenever we construe these precious aspects of our lives in an instrumental fashion—treating them as if they were impersonal products to be tallied up and bartered—it leaches the meaning from our world.

Enhancement technologies, unfortunately, are major offenders in this regard. By their very nature, they continually tend to blur the boundary between "person" and "product." The reason for this is simple: bioenhancements *are* products. If you incorporate them into your being, certain important new aspects of your body or mind will be designed to perform a specific function or achieve a particular purpose. They will be offered for sale on the open market and will come in different models, advertised and promoted by competing companies. And they will become obsolete over time, requiring periodic upgrades or replacement by newer models.

The temptation will be strong, under such circumstances, for people to start thinking of themselves—and other people—in unabashedly productlike terms. "I'm going in for an upgrade." "Sarah really does outperform Janet these days, doesn't she?" If our society reaches a point at which millions of persons are adopting bioenhancements, this psychological tendency would become real and vivid for you every time you ran up against the limits of a particular model of modification and found yourself wanting a new and improved version. It would alter the structure of your personal and professional relationships, as you experienced the incessant competition for higher levels of enhanced function. It would directly affect your socioeconomic status and your place in the ever-jostling hierarchy of society. Your enhancements, in short, would tend to commodify your being in ways that are not just ethereal and "philosophical" in nature. They would permeate the concrete reality of your lifeworld, making you feel at times like just another product struggling to keep up the pace of performance amid a seething market of rival products.

How can we fight back against commodification? One solution would be to go rigorously non-mod, but as we saw earlier, this would entail sacrifices that many individuals and families will probably be unwilling to accept. For those of us who do choose to go forward with the bioenhancement enterprise, therefore, the following strategies may prove helpful.

- *Choose enhancements with an eye to the whole person's flourishing.* In selecting a given enhancement, it is important not to fixate too closely on the specific trait that it boosts or capability that it provides, and to focus instead on how it affects the totality of your personhood and lifeworld. Instead of asking, "What does this allow me to do that I couldn't do before?" you would ask, "How does this new capability actually contribute to my overall quality of life?" Instead of asking, "What shall I do next with my newfound powers?" you would ask, "Do I really need these additional powers, and what are their indirect or hidden drawbacks?" By insistently returning to the big picture—the perspective of the whole person—you insulate yourself from the instrumental modes of valuing that commodify your being.
- *Resist the impulse to quantify your traits.* Enhancements inherently invite the quantification of features that should not be thought of as measurable. Although my strength, dexterity, and intelligence can be rated on a scale from high to low, the qualities that make up my worth as a person exist on a separate plane and cannot be approached through scalar logic.

The idea to remember here is: "I am not the same as my performance profile, and my worth does not lie in my particular capabilities but rather in the fundamental dignity of my personhood."

- *Remember the uniqueness of each person.* Comparison lies at the heart of commodification: it arises out of instrumental judgments such as, "This model is better than that one; this trait is less desirable than those." But human individuals are unique in nature and must never be compared with one another in this way. One must therefore be especially vigilant not to slip into comparative reasoning when thinking about the worth of persons.

- *Subject your enhancements to ongoing critical assessment.* Human identity is a dynamic, ever-evolving quality. As a result, features and components of your modified personhood that work harmoniously together today may acquire noxious effects over time. Consider, for example, the high-tech brain implant that you had installed six months ago. Once you figured out how to use it properly, it was amazing. You could access the Web and control machines by thought. You began communicating telepathically with people around you. Your whole sphere of experience and information suddenly expanded outward in a dramatic and exhilarating way.

But as the months have gone by, you've also begun to realize that the device is taking an unexpected toll in your life. The constant intrusion of the informational "feed" into your consciousness; the unrelenting clamor of messages and queries from the myriad machines to which you are now directly connected; the perpetual demands of attention and interaction from telepathically linked friends, acquaintances, and strangers alike—this is all starting to become a real problem.[27] Your closest relationships are suffering. You find yourself craving quiet and solitude. It dawns on you that you are serving the machines more than they are serving you.

Eventually you reach a decision. You open the implant's Settings Menu and disconnect 85 percent of the features, severely limiting the scope of the device's functioning. Your friends laugh when they hear of this and call you a Luddite. You just smile. "I made a choice," you say.

Reasserting your own values in the face of our society's headlong rush into new technologies is no easy feat. It requires setting time aside, on an ongoing basis, to assess the impact of all these frantic innovations on the things that matter most to you. But if you can pull it off, the trap of commodification is far less likely to catch you.

6. Mechanization of the self

A cup of coffee or a glass of wine (taken in moderation) exerts an impact on your psyche that is loose, unspecific, and mild: it leaves intact the basic spontaneity and unpredictability of your individual personality. With the chemicals and bioelectronics that are coming over the horizon, however, humans may well acquire a formidable instrument for modulating their own affect and mood states—an instrument that is powerful yet calibrated and precise at the same time. Unlike coffee and wine, these novel forms of modification would have the capacity to transform you into an emotional marionette whose every nuance of feeling can be orchestrated, moment by moment, from the outside—whether it is you yourself or someone else who is pulling the puppet strings. I argued in chapter 10 that these forms of self-modification could prove toxic to your humanity: the greater the control, the clearer the predictability of the outcomes—the more you risk transforming yourself into a kind of advanced humanoid AIBO.

How to avoid falling into this trap? Here are a few ideas:

- *Educate the public.* Bioenhancements that mechanize the self would probably sell like hotcakes. I call to mind the figure of Juan in my earlier story, casually popping his pills of Lighten, Ease, Wistful, Flow, and Erase. How many of us would be able to resist that level of fine-grained emotional control, particularly if the super-sophisticated chemicals deliver not only enjoyable and reliable outcomes but also negligible side effects?

  Unfortunately, the drawbacks of exercising this kind of control over ourselves would probably be subtle and incremental in nature—at least at first—and many people wouldn't even realize what kinds of harm they were doing to themselves. It will therefore be up to our educational, religious, and cultural institutions to spell out for people, in clear and persuasive terms, how high the costs of self-mechanization really are. Vigorous public awareness campaigns will no doubt be required, along the lines of today's initiatives for prevention of bullying in schools, or for enlightening the general population about the toxic effects of fatty foods and tobacco.[28]

- *Resist medicalization.* Feeling stressed? Can't sit still? Can't keep up with your workload? Any of these complaints seems sufficient nowadays to generate some doctor's prescription: "Here, take these pills and you'll be fine."[29] We need to recognize this trend as a creeping form of self-mechanization, precisely because it substitutes a chemical quick fix for the old-fashioned "character-building" practices of introspection, self-discipline, and hard work.[30] We should therefore insist that the medical profession establish more stringent guidelines that discourage the prescribing of pills as an instant solution for any and all forms of perceived mental dysfunction. If we fail to achieve this reform of everyday medical practice, the rapid advance of highly effective pharmaceuticals over the coming decades will wash over our society like a chemical tsunami. Powerful financial interests will be involved, further propelling the phenomenon along. It could be well nigh impossible to curb the incremental self-mechanization of our population if the influence and authority of the medical profession are abetting the process.

- *Beware of "moral enhancement."* The only true enhancement of morality, I argued in chapter 11, comes through a strengthening of our broad capacities for critical assessment, complex judgment, and reflective choice. Any biotechnological modification that constrains the full exercise of these faculties, subtly channeling your inclinations and decisions down preset "altruistic" paths, should be seen for exactly what it is: a partial hijacking of free will.[31] Real choice requires the ability to seriously consider the *whole* range of available alternatives, not some narrowed subset of "preapproved" options. Such modifications would constitute a deeply misguided intervention that partially mechanizes the self and degrades the very preconditions for morally meaningful action.

7. Disconnection from primary reality

Over the coming decades, advanced virtual reality devices will proffer long voyages of escape into invented universes far removed from our everyday lives. Pills and bioelectronics will allow us selectively to filter out aspects of our personal experience that we find unappealing. Our increasingly ubiquitous machines will insulate us from the physical constraints and tribulations imposed by our bodies.

All this is potentially terrific in some respects, but it can also lead us to gradually lose our bearings on reality itself. The more enthusiastically we give ourselves over to artifice, the more likely we are to drift insensibly out of touch with the planet that sustains us, the body we were born with, and the core relationships of family and community that frame our existence.[32] One already sees intimations of this problem today in the lifeworld of some city dwellers. Night and day become background abstractions if your job keeps you sealed up for eight hours in a windowless cubicle under fluorescent light. Your commute to work further reinforces this, as the subway train takes you from home to office and back through an underground warren of tunnels. Physical space becomes less real when you can interact instantly and vividly with people all over the planet. Even your food becomes a construct: the T-bone steak no longer bears much relation in your mind to the source from which it comes—a farm, a farmer, the soil, the climate, the beast that got killed in order to have its flesh sculpted into a packaged cutlet.[33]

Among the myriad writers who have sought to address this worrisome trend, the philosopher, farmer, and poet Wendell Berry stands out as one of the most persuasive. Berry's remarkable corpus of essays, novels, and poems cannot be easily classified—much less summarized—but taken as a whole they offer a thoughtful set of ideas by which each of us can help invent a different way of being within modernity.[34] Amid the framework of values that Berry celebrates, certain themes recur.

- *The place where you live.* The longer you stay there, and the more closely you observe it, the more you can learn. You gradually commence to attune your senses to the rhythms of its seasons. You are humbled by the struggles and sufferings of the creatures it sustains. You become able to experience the individual character of the land and sky around you. Even in a sprawling city like Tokyo or London, it is possible to pay attention to the earthly substrate that sustains biological life. Beneath the cracks in the sidewalks, there is dirt. Doughty weeds grow in the corners of small gardens. Insects and rodents move freely in the urban underbelly. Cultivating a keen awareness of your home place—no matter where it is—can renew your sense of rootedness in primary reality.
- *Your community.* The ever-shifting bonds that grow over time among you, your family, your neighbors, and the people in your town—all these form another ground for the cultivation of affection. To grow up alongside someone—whether a family member or a friend—is to discover those distinctive limitations and potentials of character that only reveal themselves in a currency of decades (or generations).
- *The tools you choose to use.* Whether it be a shovel or a computer, it is worth asking yourself how the device bends you to its own demands and purposes even as you take it up in service of your own aims.
- *The way you earn your daily bread, and consume it.* Your livelihood is no mere job: it embodies the elemental relationship of interdependence and reciprocity that links you to other people and to the earth beneath your

feet. The food you eat offers the most evident and direct link to the place where you live and the people in your community.   ,

- *Taking the time to be where you are*. All too often, the pressures of our work and relationships cause us to fall into patterns of headlong, unrelenting activity. It is worth making the effort to bring all this to a halt once in a while each day, taking stock, regaining a center of balance, catching your breath. Only then will it be possible to remake contact with the deeper intuitions and purposes that lend orientation to your hours and years.[35]

I have only scratched the surface here, of course. But Berry's vision does not require you to live on a rural farm, as he does, in order to benefit from it. The quietly deliberate habit he has cultivated, of getting in frontal, immediate touch with the material and social grounding of his existence, offers a potent antidote to the trend of disconnection I described above.

EIGHTEEN

What You and I Can Do Today

The first to disintegrate a nucleus was Rutherford, and there is a picture of him holding the apparatus in his lap. I then always remember the later picture when one of the famous cyclotrons was built at Berkeley, and all of the people were sitting in the lap of the cyclotron.

—Maurice Goldhaber, director, Brookhaven Laboratory (1979)[1]

Technological progress can make one feel very small and helpless at times—like a passenger on a raft, borne passively down a swift river. But this impression is an inaccurate one. Both as individuals, and when we act together in larger groups, we humans do have a say in shaping the way history unfolds. Our powers in this regard are only partial and imperfect, to be sure, but they are real, and it is a big mistake to underestimate them.[2]

As a tangible example, consider the impact of ecological ideas on our society over the past half century.[3] In the years immediately following World War II, almost no one was thinking in terms of "ecology" or "the environment." Then, in the 1960s, green ideas gradually began making inroads, taking concrete form as small, incremental victories: a new national park, a law banning certain kinds of pesticides, a recycling program, a new set of organic foods and eco-friendly products in the local supermarket. Over the decades that followed, these values gained momentum. Some industrial firms decided that cultivating a green image could be good for business. The United States created the Environmental Protection Agency; France created the Ministry of the Environment; the UN organized Earth Day and began convening regular international conferences on ecological sustainability. The founding of these new institutions, in turn, further solidified the credibility of green ideas, moving them into the mainstream and strengthening citizens' perceptions of their importance.[4] By the first decades of the new millennium, you would be hard-pressed to find a single aspect of a modern industrial democracy that remained untouched at some level by green ideas.

To be sure, our civilization still has a long road to travel if it is to achieve sustainable balance in relation to the natural world.[5] And yet, we should not underestimate the magnitude of the shift that has occurred in the fifty years since Rachel Carson published *Silent Spring*. What has taken place is not the full-scale change of course that the original environmentalists hoped for—a ninety-degree turn in a wholly new direction. Instead, what we have lived through is a more modest *deflection* of history, down a significantly different path than the trajectory that would have lain before our society if those activists had not done their work. Our civilization today is certainly not as green as the activists of the sixties wanted it to be, but it is much greener than it would have been if they had simply thrown up their hands in resignation.[6]

This story has direct implications for the shaping of bioenhancement technologies over the coming decades. We do have a say in determining what form those technologies will take, and how they will be used. If we are able to agree on at least some of the values we wish to defend, we can mobilize effectively to channel the "flow" of enhancement technologies down paths that align with those values.

THE SPACE FOR HUMAN AGENCY: POWERS AND LIMITATIONS

In a democratic polity, citizens possess three major avenues for shaping the development of science and technology: governmental action, grassroots advocacy, and self-regulation by scientists and engineers. The first of these, state action, can take many forms, including not just laws, treaties, and explicit regulations, but patents, tax policy, informal guidelines, and the funding of research. Through such administrative instruments, a nation's leaders can strongly influence what kinds of science get done, what kinds of products get designed and built, and the rate at which innovation takes place.[7] A second avenue through which citizens may shape technological advance is by direct action and advocacy at the grass roots. A classic instance of such activism occurred in the late 1950s and early 1960s, for example, when Ralph Nader took up the cause of improving automobile safety in the United States, ultimately compelling both Detroit and Washington to undertake far-reaching reforms.[8]

Scientists and engineers themselves constitute a third important factor in the governance of technological change. In February 1975, for example, a group of 140 scientists gathered at the beachside resort of Asilomar, near

Monterey, California, to discuss the implications of what was then a cutting-edge field of biotech innovation: recombinant DNA technology. Working together over several days, the participants eventually agreed on a series of voluntary ground rules regarding the biohazards of their research.[9] The conference set a remarkable precedent, proving in practice that scientists working in a potentially hazardous new field could be capable of successful self-regulation when they chose to do so.[10]

Unfortunately, these three impressive avenues for governing science and technology are counterbalanced by an equally significant array of limitations and impediments. First and foremost among these is the problem of *uncertainty*—a factor that is particularly acute when it comes to cutting-edge technologies in early stages of development. If government regulators block the development of certain potentially hazardous new technologies, for example, they can end up stifling innovation and undermining economic growth. But if they err too far in the other direction, they can end up subjecting society to potentially significant harms.[11]

A second major constraint lies in the *difficulty of reaching consensus*. Some kinds of bioethical issues, such as abortion rights or the practice of assisted suicide, are not easily amenable to compromise, because they entail trenchant either-or decisions that inevitably leave the minority side of the argument in a position of unreconcilable opposition.[12] Unfortunately, this tendency toward polarization will most likely apply to a great many issues surrounding human enhancement. Precisely because these technologies cut so deeply into the human constitution, they will tend to be perceived by the citizenry in terms of stark binary categories associated with religion or morality: good versus evil, splendid versus hideous. It will be hard for government officials to set policy over these technologies without deeply alienating significant minorities among the population.

This brings us in turn to another challenge: the problem of *informed policymaking*. As I described in early chapters, the science underlying bio-enhancement interventions is wide ranging, complicated, and subtle. If the citizens and elected leaders of tomorrow are to make intelligent choices regarding these technologies, they will need to be educated at a level that at least acquaints them with the basic parameters of many domains of specialized knowledge.[13] They will need to grapple with such subjects as the nature of genetic causation or the fundamentals of brain chemistry. If our society wishes to hold on to the Jeffersonian ideal of an informed citizenry making the key decisions regarding its own future, the schools and teachers of to-morrow will have their work cut out for them.[14]

A related problem has to do with the *bias toward short-term thinking* that characterizes human affairs.[15] Consumers tend to focus primarily on the immediate benefits conferred by a product. Elected officials serve terms that range between two and six years, and it has proved extremely difficult for them to make plans that extend over a longer time span. Unfortunately, establishing sound policies for the governance of biotechnologies requires precisely the sort of judicious, future-oriented thinking that proves hardest for our society to muster.

This is particularly problematic because of a phenomenon known as *technological lock-in*.[16] Most forms of complex technology are easiest to shape in the early stages of their development, when the basic technical parameters are still being worked out. Later on, as the technologies mature and enter mass production, it becomes much harder to modify their basic design, because by that point too many people (and business interests) have become invested in the specific form that the technology has taken. Enhancement technologies are still in their "formative years" today. They will be considerably harder to reconfigure thirty or forty years from now, when many of the basic choices regarding their design and use will already have been made.

PRACTICAL STEPS

I have tallied up several noteworthy avenues for human agency, as well as an equally impressive array of obstacles to that agency, but this does not mean that the two forces simply cancel each other out. On the contrary, even when operating amid this field of major constraints, humankind has shown a considerable ability to shape the development of science and technology over recent decades. We *can* steer the boat—partially, imperfectly, but significantly. Here, therefore, are five tangible goals people can work for as they mobilize to influence the development of human biotechnologies.

1. *Mandate basic education in science, technology, and society (STS).* It will be hard to make wise decisions about enhancement technologies unless you understand the way they work and the nature of the changes they will bring into your body and mind. A solid preparation in the basics of science and engineering—as well as their societal implications—will prove a tremendous asset in this regard. The interdisciplinary field of STS has

only existed for a few decades, but it has grown rapidly, and hundreds of universities around the world now offer courses under this rubric.[17] A wise policy would be to incorporate a requirement for such courses into the core curricula of high schools and universities. Citizens who are acquainted with essential STS concepts will be far better placed to make informed bioenhancement choices that reflect their values—whatever those values may be.[18]

2. *Build "bioethics coalitions" across the left-right divide.* Liberal and conservative intellectuals have responded to the prospect of enhancement technologies in ways that cut across existing lines of party politics: one finds progressives like Bill McKibben and Michael Sandel arguing against enhancements just as fervently as conservatives like Leon Kass or Francis Fukuyama. Pro-enhancement thinkers similarly fail to map neatly onto any left-right political alignment.[19] This provides an opening for moderates from left and right to come together in broad coalitions oriented toward finding pragmatic approaches to the governance of these technologies.[20] We should therefore urge politicians to resist the temptation to identify their side as the "pro-enhancement party" or the "non-mod party." As long as public debates remain focused on the substantive pros and cons of specific enhancement modifications—rather than on indiscriminate condemnations or celebrations of *all* enhancements—we stand a better chance of developing farsighted and effective policies.

3. *Create a strong governmental agency for technology assessment.* From 1972 to 1995, the US Congress housed an agency known as the Office of Technology Assessment (OTA), whose mission was to analyze the societal consequences of technological change—both for the present day and for the middle- and long-term future.[21] Hailed as a pathbreaking institution, whose design and functioning were subsequently emulated in many other nations, the OTA issued hundreds of in-depth reports on new technologies and their societal implications over the course of its twenty-three-year life span. It was shut down in 1995, and today some of the functions it once carried out are performed on a piecemeal basis by a variety of other government agencies.[22] But this scattershot approach is far less effective than it should be, and numerous scholars and experts have called for a new and revamped OTA to be established as soon as possible.[23] Our legislators and citizenry urgently need the kind of guidance and foresight that the OTA offered.

4. *Adopt the precautionary principle in crafting bioenhancement legislation.* The precautionary principle was first articulated by the German philosopher Hans Jonas in the 1970s, and it has now come into widespread use in drafting regulatory laws in the European Union and elsewhere.[24] It has been defined in a variety of ways, but at bottom it boils down to this: *When the suspected risks are catastrophic in nature, it is always better to err on the side of caution, even if the state of our present knowledge is imperfect.*[25] In the case of global warming, for example, the precautionary principle dictates that it is reasonable for humankind to make a significant effort (and sacrifice) today in order to avoid potentially disastrous consequences in the distant future—even though we cannot be sure precisely what those dire consequences will actually turn out to look like.[26]

As we have seen in the preceding chapters, some bioenhancement technologies will be relatively benign in their effects. One thinks, for example, of an epigenetic modification that boosts my resistance to skin cancer. Other technologies, such as fMRI devices for decoding brain activity, could prove far more Janus-faced in their implications, holding a potential for profound benefits as well as for grievous abuse. The precautionary principle can provide legislators and regulatory agencies with solid guidance for how to handle these kinds of high-stakes uncertainty.

5. *Strengthen international cooperation in governing technology.* We live in a world of ever-increasing interdependence and shared vulnerability. Unfortunately, however, the citizens and elected leaders of most countries still think of science and technology as prime *national* assets in the relentless global competition for power and resources. As long as this attitude prevails, the international governance of technology will remain a weak link in humankind's ability to shape its own future. The only solution here, to put it bluntly, is for more and more people to wake up and smell the coffee. The days of national prowess and prosperity—considered as something separate from the flourishing of global civilization—are already over. Citizens who hold up placards saying "USA first!" or "China first!" may think they are defending a sacred patriotic interest, but in fact they are harming both themselves and their beloved country. Such narrow nationalisms will become more dramatically obsolete and counterproductive with every passing year.[27]

Seen in this light, the creation of a global OTA would constitute an excellent step. It would provide an instrument not only for coordinating technological innovation and regulation, but for grappling with planet-level challenges related to ecological, economic, medical, and security-related problems as well. Unless we citizens demand it, our governments will proceed by thimblefuls in making the necessary reforms and building the requisite international institutions. We should insist on bucketfuls of change instead.

THE RIGHT MEASURE OF REALISM AND IDEALISM

At one level, the strategies I have been describing in the past three chapters lean heavily toward idealism. A global Marshall Plan? A planetwide moratorium on radical forms of self-modification? A citizenry that educates itself about the dehumanizing aspects of these technologies and exercises restraint in adopting them? For someone who has studied history even a little bit, these proposals cannot help but seem excessively optimistic. They may be valid in principle, but what are their chances of ever becoming a reality?

This is a reasonable objection, but it is susceptible to an equally reasonable answer. Just because something that we value highly may be impossible to achieve in its entirety, this does not mean that we should simply give up on it. It is possible (and important) to work toward worthy goals, even if we know that we may never fully realize them. Here I return to the example of the green movement. If the ecological activists of the 1960s like Rachel Carson and E. F. Schumacher had adopted a rigorously "realistic" assessment of the forces arrayed against them, they would have had good reason for despair: the entire structure of the global economy was oriented toward precisely the sort of headlong, mindless growth that they were criticizing. Their views were only shared by a small minority, and the overall society and culture around them tended to dismiss them as fringe radicals. "Get real," they were told.

But they did not let this deter them. They published articles and books, they organized meetings and rallies, they formed coalitions, and slowly, incrementally, persuaded larger and larger numbers of people with their ideas. Eventually, over several decades, they succeeded in partially—but significantly—altering the trajectory of modern history. To be "idealistic," in this sense, is actually a potent form of realism: it is a strategy that aims at gradual, long-term change, oriented by values it holds dear.

The same principle applies to the values we choose to assert—as individuals, as families, as a society—with regard to the enhancement enterprise. If we believe, for example, that equality of opportunity is a good thing, then we should do what we can to defend it and to prevent a biostratified caste system from forming. The mere fact that socioeconomic inequality is a deep and intractable feature of human history should not deter us. Our efforts today can open up new pathways and possibilities for our children and grandchildren—pathways that they in turn can expand and develop more fully, in ways that we cannot even imagine.

NINETEEN

Enhancing Humility

Some Concluding Thoughts

Anche l'ultima città dell'imperfezione ha la sua ora perfetta, l'ora, l'attimo, in cui in ogni città c'è la Città. (Even the last city of imperfection has its perfect hour, the hour, the instant, when in every city there is the City.)

—Italo Calvino, *La giornata d'uno scrutatore* (1963)[1]

ALBERT EINSTEIN AND LEO SZILARD

Leo Szilard arrived at the Long Island summer house of Albert Einstein in the mid-afternoon of July 16, 1939.[2] His chauffeur that day was a fellow physicist, Eugene Wigner (Nobel Prize, 1963). Einstein knew both men well—Szilard had been his student back in Berlin during the early 1920s. Szilard did not beat around the bush. He shared with his old professor the latest news in nuclear physics, especially the recent discoveries made by scientists in Nazi Germany. Then he laid out for Einstein the concept of the nuclear chain reaction, an idea that Szilard had first conceived in 1933, then kept secret because it frightened him so deeply. Einstein looked amazed, and exclaimed, "*Daran habe ich gar nicht gedacht!*"—"I never thought of that!"[3]

A few hours later, Szilard and Wigner drove back to Manhattan, bearing the letter they had persuaded Einstein to sign. It was addressed to President Roosevelt, and it detailed the possibility that the Germans might be on their way to building an atomic bomb: "It is conceivable—though much less certain—that extremely powerful bombs of a new type may thus be constructed. A single bomb of this type, carried by boat and exploded in a port, might very well destroy the whole port together with some of the surrounding territory."[4]

Out of these events the Manhattan Project eventually materialized, with Szilard and Wigner working alongside Enrico Fermi in Chicago. Einstein, a lifelong pacifist, refused to take part. After the war, he described his signing of Szilard's letter as the single greatest mistake of his life.[5]

Szilard, too, agonized after 1945 about the implications of what he had done. He realized that the possibility of a Hitler wielding atomic weapons had justified taking extreme countermeasures. But he felt deeply pessimistic about the long-term ability of humankind to deal with a technology of such destructive power. He died in 1964, still tortured by the thought that he had handed to his species the instrument of its own extinction.

Nuclear technology, in short, was born in a weapons-ready mode. Although scientists had speculated from the start about harnessing this elemental fire for civilian uses, in actual practice it came into being with the purpose of mass killing. Szilard and Einstein both knew this all too well. They mourned the strange historical coincidence that had brought such a godlike instrument into human affairs in a century when our civilization was so obviously unprepared to handle it.

Enhancement technologies are profoundly different, in this regard. They have grown primarily out of medicine, biological science, and bioengineering—our generous impulses to heal the sick and to improve the quality of human lives. Though military agencies such as DARPA have certainly shown keen interest in promoting the development of superhuman capabilities, most of the pharmaceuticals, bioelectronics, and genetic interventions we have discussed in this book have come into being through the civilian world. Even two such visionaries as Szilard and Einstein could not have foreseen this irony. A mere seven decades after Hiroshima, our science and technology have grown so powerful that even their most benign applications now stand poised to turn our civilization upside down. Our instruments of healing hold the potential of destabilizing our future just as profoundly as our instruments of destruction.

Nuclear weaponry and bioenhancement technologies do have one important feature in common, however: their discovery cannot be undone. Like the power of the atom, enhancement technologies are with us now for good. We will have to learn how to live with them.

■ ■ ■

When you acquire a godlike power (and you are not a god), an excellent virtue to cultivate is that of humility. Here are two forms of humility that will stand our species in good stead over the coming century.

HUMILITY IN THE PACING OF SOCIETAL CHANGE

Precisely because I am a historian, I am usually loath to speak of the lessons of history. The past has too many conflicting lessons embedded in it to be of much use as a general guidebook for orienting our choices. Chances are, if you are seeking practical wisdom in the past, you will (whether knowingly or unwittingly) cherry-pick precisely those events that reinforce the precepts you were seeking to advocate in the first place. Someone with a different agenda could no doubt comb through the long record and build a fine case for the opposite conclusions.

Having said this, however, I am now going to hazard precisely such a "lesson," and what's worse, I will articulate it in the form of a sweeping generalization. *Reform works better than revolution.* Strategies of slow, incremental change have succeeded far better at achieving the aims of historical actors than strategies of sudden, drastic change. I make this claim in the full knowledge that it can be qualified in all sorts of significant ways.[6] Nevertheless, it leaps out at me from the mass of historical events with such intuitive force that I feel compelled to take it seriously. I bring it up here because it has major implications for how our society chooses to pursue the bioenhancement enterprise over the coming century.

Consider the three major revolutionary episodes of the modern era: 1789, 1917, 1949. These experiments with sudden radical transformation certainly brought about far-reaching impacts for the societies in which they occurred. But all three skidded eventually out of control, ultimately failing to realize the intentions of the people who had launched them. The political and civic revolutionaries in Paris ended up on the guillotine or under the iron rule of Napoleon. The Marxist ideals of 1917 became a bizarre Orwellian nightmare under Stalin. The Maoist principles of 1949 reached their apotheosis in the famine of 1958–1962 and vicious factional strife of the Cultural Revolution. I am not arguing here that these three great turning points did not also generate *some* significant positive effects, both directly and (especially) indirectly. Rather, I am underscoring the fact that, on balance, they failed to realize the goals of the men and women who set them into motion, and ultimately led to disastrous outcomes that shattered millions of lives.[7]

If one compares these titanic upheavals with three major reformist movements that took place during the same era, the contrast is striking.[8] The campaign in the West for equal rights for women has unfolded over two centuries, beginning in the early 1800s. Eschewing violent methods,

and adopting instead a tenacious strategy of incremental inroads and reforms, women have succeeded over a dozen generations in utterly transforming their status and power in the social order. This is not to say that full
equality has been achieved yet, but if one compares the position of women
in 1815 with where it stands today, the difference is breathtaking.[9] A similar strategy—and similar success—have characterized those portions of the
working-class movement that rejected revolutionary methods and embraced
gradual reform instead. Over the same two centuries, their rights and power
in Western democracies have steadily increased, as trade unions, the vote,
public education, and direct political influence have slowly transformed their
socioeconomic status. Again, the victory is not absolute or complete, but the
contrast with the era of Charles Dickens could not be starker.[10] Finally, the
position of blacks in America offers yet another vivid example of reformist
success. From the appalling conditions of the Reconstruction period to the
presidency of Barack Obama, the change process has been long and hard,
but a strategy of unrelenting, nonviolent pressure, aiming at one incremental goal after another, has gradually transformed the lifeworld of African
Americans. Much remains to be achieved, but young blacks today face a
dramatically broader universe of possibilities than their great-grandparents
did.[11] Gradual reform, in short, is not just morally superior because of its
generally nonviolent character: it is also more *effective* in the long run, en-
gendering forms of enduring change that penetrate deeply into the fabric of
society, altering hearts and minds as well as institutions.[12]

When it comes to the pursuit of the enhancement enterprise, therefore,
our society would do well to take the comparative history of reform and revolution into account. We should choose the long, slow, plodding road rather
than the shining superhighway of radical change. Technological innovation
may indeed be accelerating, but we should not allow it to transform our lives
more rapidly than our social, cultural, and moral frameworks can absorb. If
we permit enhancement technologies to advance too quickly, the resultant
stresses could end up massively destabilizing our civilization, perhaps even
tearing it apart.

I place particular emphasis here on the idea of a phased moratorium—
moving incrementally over many decades (and perhaps centuries) from
relatively low levels of enhancement toward deeper and more radical modifications. Despite all the practical challenges and moral dilemmas that such
a policy of restraint would entail, it ultimately makes more sense to pursue this cautious path of structured change than to plunge headlong into a

species-level experiment of unlimited bioenhancement. No one can know today how long this process of voluntarily restrained change will need to be drawn out: it is something that the coming generations will have to gauge for themselves—and debate among themselves—on an ongoing basis. But the underlying principle—*go slow enough to let your social order adapt*—meshes well with both the historical record and plain common sense.

HUMILITY IN WHAT WE EXPECT FROM TECHNOLOGY

If you have gotten a chance in recent years to sit at the back of a large class-room during a lecture in a college course, you will perhaps have had the same unnerving experience as I. The lecture hall spreads before you like a sloping amphitheater, the professor a tiny gesticulating figure way down there. You stare out over a sea of the backs of heads. Every student has his or her laptop open: the screens glow in the semi-darkened room. The profes-sor walks back and forth, gesturing, explaining some concept.

You notice that the person in front of you is checking her e-mail. A quick glance to the right: the next student is on an airline website, intent on book-ing a flight. Glance to the left: that student is playing computer solitaire. You frown, and shift your attention from the lecturer. How many of these kids are actually paying attention to the class? You peer over the rows of glowing laptop screens. A few of them seem open to word-processing applications: those students are taking notes. But the more you gaze around the room, the more it hits you. At least two-thirds of the screens are alive with the movement of games, e-mail, Facebook, YouTube, the Web—anything but the classroom.

The professor walking back and forth down there is a nationally re-spected scholar who has also won teaching awards. This is a highly selective university that attracts some of the ablest students in the world. The lecture the professor has prepared is subtle, well crafted, and fascinating. The topic is vitally important. He is delivering it in a vivid, engaging manner, replete with occasional photos and video clips. And yet the majority of persons in this room are paying scant attention to him. Am I the only one in here who's thinking, "WTF?"

I present this story because it exemplifies what happens when people expect something from technology that it can't actually deliver. Those of us in higher education have shown ourselves to be particularly gullible in this regard. As new computer tools were developed over the past twenty years, we eagerly got on the bandwagon: we converted our lectures to

PowerPoint, incorporated online forums and handheld interactive devices into our courses, badgered our administrators to pay for the fanciest "smart classrooms." As students started taking their laptops to class, we assumed as a matter of course that this would facilitate more detailed note taking.

But we failed, all too often, to pose the key underlying questions. Are all these newfangled devices genuinely increasing the quality of interaction among students and teachers? Is the learning that takes place getting any deeper or more incisive?[13] The thought that—gasp—this armamentarium of gadgetry might actually be *undermining* our educational goals seemed unimaginable.

I am not merely making a case here for a more critical assessment of technology, and for selectively reining it in. I am also arguing for a sober understanding of what it can and cannot deliver. Technology is a means, never an end in itself. Our contemporary society—especially in the West, but increasingly in other parts of the world as well—seems to have forgotten this basic fact. Lured on by the talismans of "new" and "more," we have too often lost sight of the aims that our tools and instruments were intended to serve. We have become dazzled by the hammer, forgetting that our purpose was to drive nails.

Technology can give you greater capabilities, but it will not tell you what to do with them. It can augment your senses, but only you will be the one to integrate those novel experiences into a meaningful life narrative. It can increase your connection with other people, but those connections will only have value insofar as you sincerely invest yourself in them, opening up to those around you. These are truisms—indeed, clichés—but the headlong rush of our modern lives can lead us to lose touch with the underlying message they convey. A great deal of contemporary culture encourages us to fixate on things outside us, ahead of us, believing that if only we can obtain them, we will find an enduring happiness. And yet, as we must invariably discover, that is a mirage that continually dances away.[14]

Here, therefore, lies a second form of hubris: the notion that we can live much better lives simply by boosting our traits and increasing our powers. This is an illusion. New machines, proficiencies, capabilities—these are only worth pursuing insofar as they serve the core qualities that carry intrinsic value: moments of fellowship; journeys of discovery; a glimmering of insight; the breakthrough of forgiveness; a flash of humor; getting to understand each other; creating a thing of beauty; commiserating, helping each other out, holding each other tight. Technology in itself cannot deliver these, nor do we require technology in order to attain them. They do not lie out there, around the corner of tomorrow. They are here, ready to hand, inside each of us.

Over the coming century, Homo sapiens may well become ever more powerful and smart and dazzling with sheer radiant capability. That is the road down which our civilization is eagerly headed. If we play our cards right, we may yet be able to avoid the nastier scenarios I have described in the preceding chapters. We may succeed in building an exhilarating, fascinating future for ourselves.

But all this will prove a blind and futile flailing, if we lose sight of the qualities that bring fulfillment into a human life. These qualities do not await us in the future. They have walked alongside us throughout our history—embodied not just in famous figures like St. Francis of Assisi or Aung San Suu Kyi, but in all the nameless individuals who have given themselves over to others, and who continue to do so every day in the community that surrounds us. Each of us knows such persons—the quiet givers. In that future world, as in the world of today, the most potent deed of all will still take form as a smile, a silent nod of empathy, a hand gently laid on someone's arm. The merest act of kindness will still remain the Ultimate Enhancement.

SELECTED FILMOGRAPHY

Note: For updates to this list, see this book's companion website at www.ourgrandchildrenredesigned.org.

2001

2010

AI: Artificial Intelligence

Alien Resurrection

Avatar

Battlestar Galactica (TV series)

Bicentennial Man

Blade Runner

Brainstorm

Cocoon

Deus Ex (video game)

Flowers for Algernon

Frankenstein (1994)

Gattaca

Her

I, Robot

Iron Man

Lawnmower Man

Limitless

The Matrix

"The Measure of a Man" (episode of *Star Trek: The Next Generation*)

Metropolis

Minority Report

Moon

Never Let Me Go

Rise of the Planet of the Apes

Robocop

Spider-Man (2002)

Splice

Star Trek (all the TV series)

Star Wars

The Six Million Dollar Man (TV series)

Source Code

Strange Days

Surrogates

Terminator (all the films)

Terminal Man

Total Recall

Transcendent Man (documentary)

Wall-e

SELECTED BIBLIOGRAPHY / FURTHER READING

Note: For the full bibliography, see this book's companion website at www.ourgrandchildrenredesigned.org.

This bibliography's thematic headings are

- Animal enhancement
- Bioelectronic enhancement and cyborgs
- Brain/mind/neuroethics
- Cosmetics, life extension, and physical enhancements
- Enhancement and posthumanity: bioethical and social implications
- Forecasting and futurology
- Genetics and genetic enhancement
- Human nature, human identity, human dignity, disability studies
- Justice, ethics
- Law, government policy, military applications
- Nanotechnology, virtual reality, synthetic biology
- Pharmaceutical enhancement
- Robots and artificial intelligence
- Science fiction
- Science studies, history of technology, technological determinism

Animal enhancement

Anthes, Emily. *Frankenstein's Cat: Cuddling Up to Biotech's Brave New Beasts.* New York: Scientific American/Farrar, Straus, and Giroux, 2013.

Beauchamp, Tom, and R. G. Frey, eds. *The Oxford Handbook of Animal Ethics.* New York: Oxford University Press, 2011.

Bonnicksen, Andrea. *Chimeras, Hybrids, and Interspecies Research: Politics and Policymaking.* Washington, DC: Georgetown University Press, 2009.

Haraway, Donna. *When Species Meet*. Minneapolis: University of Minnesota Press, 2008.

Hauser, Marc. *Wild Minds: What Animals Really Think*. New York: Holt, 2000.

Hess, Elizabeth. *Nim Chimpsky: The Chimp Who Would Be Human*. New York: Bantam, 2008.

Regan, Tom. *The Case for Animal Rights*. Oakland: University of California Press, 2004.

Rollin, Bernard. *The Frankenstein Syndrome: Ethical and Social Issues in the Genetic Engineering of Animals*. New York: Cambridge University Press, 1995.

Sunstein, Cass, and Martha C. Nussbaum, eds. *Animal Rights: Current Debates and New Directions*. New York: Oxford University Press, 2004.

Bioelectronic enhancement and cyborgs

Belfiore, Michael. *The Department of Mad Scientists: How DARPA Is Remaking Our World, from the Internet to Artificial Limbs*. New York: Harper, 2009.

Berger, Theodore, et al. *Brain-Computer Interfaces: An International Assessment of Research and Development Trends*. New York: Springer, 2008.

Chorost, Michael. *Rebuilt: How Becoming Part Computer Made Me More Human*. Boston: Houghton Mifflin, 2005.

Clark, Andy. *Natural-Born Cyborgs: Minds, Technologies, and the Future of Human Intelligence*. New York: Oxford University Press, 2003.

Haraway, Donna. "A Cyborg Manifesto: Science, Technology, and Socialist-Feminism in the Late-Twentieth Century." In *Simians, Cyborgs, and Women: The Reinvention of Nature*, 149–81. New York: Routledge, 1991.

Hughes, James. *Citizen Cyborg: Why Democratic Societies Must Respond to the Redesigned Human of the Future*. New York: Westview, 2004.

Katz, Bruce. *Neuroengineering the Future: Virtual Minds and the Creation of Immortality*. Sudbury, MA: Infinity Science Press, 2008.

Nicolelis, Miguel. *Beyond Boundaries: The New Neuroscience of Connecting Brains With Machines—and How It Will Change Our Lives*. New York: Times Books, 2011.

Rose, Steven. *The Future of the Brain: The Promise and Perils of Tomorrow's Neuroscience*. New York: Oxford University Press, 2005.

Wolpaw, Jonathan, and Elizabeth Wolpaw, eds. *Brain-Computer Interfaces: Principles and Practice*. New York: Oxford University Press, 2012.

Brain/mind/neuroethics

Baumeister, Alan. "The Tulane Electrical Brain Stimulation Program: A Historical Case Study in Medical Ethics." *Journal of the History of the Neurosciences* 9, no. 3 (2000): 262–78.

Bedau, Mark, and Paul Humphreys, eds. *Emergence: Contemporary Readings in Philosophy and Science*. Cambridge, MA: MIT Press, 2008.

Berger, Theodore, and Dennis Glanzman. *Toward Replacement Parts for the Brain: Implantable Biomimetic Electronics as Neural Prostheses*. Cambridge, MA: MIT Press, 2005.

Chalmers, David, ed. *Philosophy of Mind: Classical and Contemporary Readings*. New York: Oxford University Press, 2002.

Churchland, Patricia. *Braintrust: What Neuroscience Tells Us About Morality*. Princeton, NJ: Princeton University Press, 2011.

Damasio, Antonio. *The Feeling of What Happens: Body and Emotion in the Making of Consciousness*. New York: Harcourt, 1999.

Dehaene, Stanislas. *Reading in the Brain*. New York: Penguin, 2009.

Delgado, José. *Physical Control of the Mind: Toward a Psychocivilized Society*. New York: Harper, 1969.

Dennett, Daniel C. *Consciousness Explained*. Boston: Little, Brown, 1991.

Edelman, Gerald M. *Second Nature: Brain Science and Human Knowledge*. New Haven, CT: Yale University Press, 2006.

Gazzaniga, Michael. *Who's in Charge? Free Will and the Science of the Brain* New York: Harper Collins, 2011.

Giordano, James, and Bert Gordijn, eds. *Scientific and Philosophical Perspectives in Neuroethics*. New York: Cambridge University Press, 2010.

Glannon, Walter, ed. *Defining Right and Wrong in Brain Science: Essential Readings in Neuroethics*. New York: Dana, 2007.

Hawkins, Jeff. *On Intelligence*. New York: New York Times Books, 2004.

Hofstadter, Douglas. *I Am a Strange Loop*. New York: Basic, 2007.

Holland, John. *Emergence: From Chaos to Order*. New York: Basic, 1998.

Humphrey, Nicholas. *Soul Dust: The Magic of Consciousness*. (Princeton, NJ: Princeton University Press, 2011.

Illes, Judy, ed. *Neuroethics: Defining the Issues in Theory, Practice, and Policy*. New York: Oxford University Press, 2006.

Juarrero, Alicia. *Dynamics in Action: Intentional Behavior as a Complex System*. Cambridge, MA: MIT Press, 1999.

Kandel, Eric R. *In Search of Memory: The Emergence of a New Science*. New York: Norton, 2006.

Kane, Robert, ed. *The Oxford Handbook of Free Will*, 2nd ed. New York: Oxford University Press, 2011.

Koch, Christof. *The Quest for Consciousness: A Neurobiological Approach*. Greenwood Village, CO: Roberts, 2004.

Lakoff, George, and Mark Johnson. *Philosophy in the Flesh: The Embodied Mind and its Challenge to Western Thought*. New York: Basic, 1999.

LeDoux, Joseph. *Synaptic Self: How Our Brains Become Who We Are*. New York: Viking, 2002.

Levy, Neil. *Neuroethics: Challenges for the 21st Century*. New York: Cambridge University Press, 2007.

Metzinger, Thomas. *The Ego Tunnel: The Science of the Mind and the Myth of the Self*. New York: Basic, 2009.

Minsky, Marvin. *The Emotion Machine: Commonsense Thinking, Artificial Intelligence, and the Future of the Human Mind*. New York: Simon & Schuster, 2006.

Moreno, Jonathan. *Mind Wars: Brain Research and National Defense*. New York: Dana, 2006.

Searle, John. *The Mystery of Consciousness*. New York: New York Review of Books, 1997.

Seung, Sebastian. *Connectome: How the Brain's Wiring Makes Us Who We Are*. New York: Houghton Mifflin Harcourt, 2012.

Zak, Paul. *The Moral Molecule: The Source of Love and Prosperity*. New York: Dutton, 2012.

Cosmetics, life extension, and physical enhancements

Arrison, Sonia. *100 Plus: How the Coming Age of Longevity Will Change Everything, From Careers and Relationships to Family and Faith*. New York: Basic, 2011.

De Grey, Aubrey. *Ending Aging: The Rejuvenation Breakthroughs that Could Reverse Human Aging in Our Lifetime*. New York: St. Martin's, 2007.

Hall, Stephen S. *Merchants of Immortality: Chasing the Dream of Human Life Extension*. New York: Houghton Mifflin, 2003.

Hoberman, John. *Testosterone Dreams: Rejuvenation, Aphrodisia, Doping*. Oakland: University of California Press, 2005.

Immortality Institute. *The Scientific Conquest of Death: Essays on Infinite Lifespans*. Immortality Institute, 2004.

Jendrick, Nathan. *Dunks, Doubles, Doping: How Steroids Are Killing American Athletics*. Guilford, CT: Lyons, 2006.

Magary, Drew. *The Postmortal*. New York: Penguin, 2011.

Medina, John. *The Clock of Ages: Why We Age, How We Age, Winding Back the Clock*. New York: Cambridge University Press, 1996.

Overall, Christine. *Aging, Death, and Human Longevity: A Philosophical Inquiry*. Oakland: University of California Press, 2003.

Post, Steven, and Robert Binstock, eds. *The Fountain of Youth: Cultural, Scientific, and Ethical Perspectives on a Biomedical Goal*. New York: Oxford University Press, 2004.

Stipp, David. *The Youth Pill: Scientists at the Brink of an Anti-Aging Revolution*. New York: Current Trade, 2013.

Tamburrini, Claudio, and Torbjorn Tännsjo, eds. *Genetic Technology and Sport: Ethical Questions*. New York: Routledge, 2005.

Zey, Michael. *Ageless Nation: The Quest for Superlongevity and Physical Perfection*. Far Hills, NJ: New Horizon Press, 2007.

Enhancement and posthumanity: bioethical and social implications

Agar, Nicholas. *Humanity's End: Why We Should Reject Radical Enhancement*. Cambridge, MA: MIT Press, 2010.

Agar, Nicholas. *Liberal Eugenics: In Defence of Human Enhancement*. Malden, MA: Blackwell, 2004.

Buchanan, Allen. *Better Than Human: The Promise and Perils of Enhancing Ourselves*. New York: Oxford University Press, 2011.

Buchanan, Allen. *Beyond Humanity? The Ethics of Biomedical Enhancement*. New York: Oxford University Press, 2011.

Fukuyama, Francis. *Our Posthuman Future: Consequences of the Biotechnology Revolution*. New York: Farrar, Straus, and Giroux, 2002.

Habermas, Jürgen. *The Future of Human Nature*. Cambridge, UK: Polity, 2003.

Harris, John. *Enhancing Evolution: The Ethical Case for Making Better People*. Princeton, NJ: Princeton University Press, 2007.

Joy, Bill. "Why the Future Doesn't Need Us." *Wired*, April 2000.

Kass, Leon. *Beyond Therapy: Biotechnology and the Pursuit of Happiness*. New York: Regan, 2003.

Kurzweil, Ray. *The Singularity Is Near: When Humans Transcend Biology*. New York: Viking, 2005.

Lanier, Jaron. *You Are Not a Gadget: A Manifesto.* New York: Knopf, 2010.

McKibben, Bill. *Enough: Staying Human in an Engineered Age.* New York: Times Books, 2003.

Naam, Ramez. *More Than Human: Embracing the Promise of Biological Enhancement.* New York: Broadway, 2005.

Parens, Erik, ed. *Enhancing Human Traits: Ethical and Social Implications.* Washington, DC: Georgetown University Press, 1998.

Pence, Gregory. *How to Build a Better Human: An Ethical Blueprint.* Lanham, MD: Rowman & Littlefield, 2012.

Persson, Ingmar, and Julian Savulescu. *Unfit for the Future: The Need for Moral Enhancement.* New York: Oxford University Press, 2012.

Sandel, Michael. *The Case Against Perfection: Ethics in the Age of Genetic Engineering.* Cambridge, MA: Harvard University Press, 2007.

Savulescu, Julian, Ruud ter Meulen, and Guy Kahane, eds. *Enhancing Human Capacities.* Hoboken, NJ: Wiley-Blackwell, 2011.

Savulescu, Julian, and Nick Bostrom, eds. *Human Enhancement.* New York: Oxford University Press, 2009.

Forecasting and futurology

Barr, Marleen, ed. *Envisioning the Future: Science Fiction and the Next Millennium.* Middletown, CT: Wesleyan University Press, 2003.

Bell, Wendell. *Foundations of Futures Studies: Human Science for a New Era.* Volume 1. Piscataway, NJ: Transaction Publishers, 1997.

Clarke, Arthur. *Profiles of the Future: An Inquiry into the Limits of the Possible.* London: Indigo, 1999.

Dyson, Freeman. *Imagined Worlds.* Cambridge, MA: Harvard University Press, 1997.

Gore, Al. *The Future: Six Drivers of Global Change.* New York: Random House, 2013.

Lombardo, Thomas. *Contemporary Futurist Thought.* Bloomington, IN: Authorhouse, 2006.

Seidensticker, Bob. *Futurehype: The Myths of Technology Change.* San Francisco: Berrett-Koehler, 2006.

Genetics and genetic enhancement

Barnes, Barry, and John Dupré. *Genomes: And What to Make of Them.* Chicago: University of Chicago Press, 2008.

Black, Edwin. *War Against the Weak: Eugenics and America's Campaign to Create a Master Race.* 2nd ed. Washington, DC: Dialog, 2012.

Buchanan, Allen, Dan W. Brock, Norman Daniels, and Daniel Wikler. *From Chance to Choice: Genetics and Justice.* New York: Cambridge University Press, 2000.

Burley, Justine, and John Harris, eds. *A Companion to Genethics.* Malden, MA: Blackwell, 2002.

Carey, Nessa. *The Epigenetics Revolution: How Modern Biology Is Rewriting Our Understanding of Genetics, Disease, and Inheritance.* New York: Columbia University Press, 2012.

DeGrazia, David, *Creation Ethics: Reproduction, Genetics, and Quality of Life.* New York: Oxford University Press, 2012.

Francis, Richard. *Epigenetics: How Environment Shapes Our Genes*. New York: Norton, 2011.

Green, Ronald M. *Babies by Design: The Ethics of Genetic Choice*. New Haven, CT: Yale University Press, 2007.

Heyd, David. *Genethics: Moral Issues in the Creation of People*. Oakland: University of California Press, 1992.

Jablonka, Eva, and Marion Lamb. *Evolution in Four Dimensions: Genetic, Epigenetic, Behavioral, and Symbolic Variation in the History of Life*. Cambridge, MA: MIT Press, 2005.

Kevles, Daniel. *In the Name of Eugenics: Genetics and the Uses of Human Heredity*. Cambridge, MA: Harvard University Press, 2004.

McGee, Glenn. *The Perfect Baby: Parenthood in the New World of Cloning and Genetics*. Lanham, MD: Rowman & Littlefield, 2000.

Mehlman, Maxwell. *Transhumanist Dreams and Dystopian Nightmares: The Promise and Peril of Genetic Engineering*. Baltimore: Johns Hopkins University Press, 2012.

Parens, Erik, Audrey Chapman, and Nancy Press, eds. *Wrestling with Behavioral Genetics: Science, Ethics, and Public Conversation*. Baltimore: Johns Hopkins University Press, 2006.

Ridley, Matt. *Nature via Nurture: Genes, Experience, and What Makes Us Human*. New York: Harper, 2003.

Silver, Lee M. *Remaking Eden: How Genetic Engineering Will Transform the American Family*. New York: Harper, 1998.

Stock, Gregory. *Redesigning Humans: Choosing Our Genes, Changing Our Future*. Boston: Mariner, 2003.

Human nature, human identity, human dignity, disability studies

Albrecht, Gary, et al. *Handbook of Disability Studies*. Thousand Oaks, CA: Sage, 2001.

Arnhart, Larry. *Darwinian Natural Right: The Biological Ethics of Human Nature*. Albany: State University of New York Press, 1998.

Berry, Wendell. *Life Is a Miracle: An Essay Against Modern Superstition*. Washington, DC: Counterpoint, 2000.

Brown, Donald. *Human Universals*. New York: McGraw-Hill, 1991.

DeGrazia, David. *Human Identity and Bioethics*. New York: Cambridge University Press, 2005.

Douglas, Mary. *Purity and Danger: An Analysis of Concepts of Pollution and Taboo*. New York: Routledge, 1966.

Dupré, John. *Human Nature and the Limits of Science*. New York: Oxford University Press, 2001.

Gazzaniga, Michael S. *Human: The Science Behind What Makes Us Unique*. New York: Harper Collins, 2008.

Glover, Jonathan. *What Sort of People Should There Be?* New York: Penguin, 1984.

Gilbert, Daniel. *Stumbling on Happiness*. New York: Knopf, 2007.

Haidt, Jonathan. *The Happiness Hypothesis: Finding Modern Truth in Ancient Wisdom*. New York: Basic, 2006.

Kateb, George. *Human Dignity*. Cambridge, MA: Harvard University Press, 2011.

Nussbaum, Martha. *Creating Capabilities: The Human Development Approach*. Cambridge, MA: Harvard University Press, 2011.

Peterson, Christopher, and Martin Seligman. *Character Strengths and Virtues: A Handbook and Classification.* New York: Oxford University Press, 2004.

Pinker, Steven. *The Blank Slate: The Modern Denial of Human Nature.* New York: Viking, 2002.

Smith, Christian. *What Is a Person? Rethinking Humanity, Social Life, and the Moral Good from the Person Up.* Chicago: University of Chicago Press, 2010.

Sowell, Thomas. *A Conflict of Visions: Ideological Origins of Political Struggles.* New York: Basic, 2007.

Stevenson, Leslie, and David Haberman. *Ten Theories of Human Nature.* 5th ed. New York: Oxford University Press, 2009.

Wright, Robert. *The Moral Animal: Why We Are the Way We Are.* New York: Vintage, 1994.

Justice, ethics

Copp, David, ed. *The Oxford Handbook of Ethical Theory.* New York: Oxford University Press, 2006.

Dworkin, Ronald. *Justice for Hedgehogs.* Cambridge, MA: Harvard University Press, 2011.

Kitcher, Philip. *The Ethical Project.* Cambridge, MA: Harvard University Press, 2011.

LaFollette, Hugh, ed. *The Blackwell Guide to Ethical Theory.* Malden, MA: Blackwell, 2000.

Lemos, Noah. *Intrinsic Value: Concept and Warrant.* New York: Cambridge University Press, 2009.

MacIntyre, Alasdair. *Dependent Rational Animals: Why Human Beings Need the Virtues.* Chicago: Open Court, 1999.

Rawls, John, and Erin Kelly. *Justice as Fairness: A Restatement.* 2nd ed. Cambridge, MA: Belknap, 2001.

Sandel, Michael. *Justice: What's the Right Thing to Do?* New York: Farrar, Straus, and Giroux, 2009.

Sandel, Michael. *What Money Can't Buy: The Moral Limits of Markets.* New York: Farrar, Straus, and Giroux, 2012.

Satz, Debra. *Why Some Things Should Not Be for Sale: The Moral Limits of Markets.* New York: Oxford University Press, 2010.

Sen, Amartya. *The Idea of Justice.* Cambridge, MA: Belknap, 2009.

Singer, Peter. *Practical Ethics.* 3rd ed. Cambridge, UK: Cambridge University Press, 2011.

Sunstein, Cass. *Laws of Fear: Beyond the Precautionary Principle.* New York: Cambridge University Press, 2005.

Law, government policy, military applications

Boulding, Kenneth. *Stable Peace.* Austin: University of Texas Press, 1978.

Brin, David. *The Transparent Society: Will Technology Force Us to Choose Between Privacy and Freedom?* New York: Basic, 1998.

Coyle, Diane. *The Economics of Enough: How to Run the Economy as if the Future Matters.* Princeton, NJ: Princeton University Press, 2011.

ETC Group. *Who Owns Nature? Corporate Power and the Final Frontier in the Commodification of Life*. Ottawa, Canada: ETC, November 2008.

Jonas, Hans. *The Imperative of Responsibility: In Search of an Ethics for the Technological Age*. Chicago: University of Chicago Press, 1985.

Jotterrand, Fabrice, ed. *Emerging Conceptual, Ethical, and Policy Issues in Bionanotechnology*. New York: Springer, 2008.

Koblenz, Gregory. *Living Weapons: Biological Warfare and International Security*. Ithaca, NY: Cornell University Press, 2011.

Krimsky, Sheldon, and Peter Shorett, eds. *Rights and Liberties in the Biotech Age: Why We Need a Genetic Bill of Rights*. Lanham, MD: Rowman & Littlefield, 2005.

Levinson, Bradley, and Doyle Stevick. *Reimagining Civic Education: How Diverse Societies Form Democratic Citizens*. Lanham, MD: Rowman & Littlefield, 2007.

Marchant, Gary, Braden Allenby, and Joseph Herkert, eds. *The Growing Gap Between Emerging Technologies and Legal-Ethical Oversight*. New York: Springer, 2011.

Sklove, Richard. *Reinventing Technology Assessment: A 21st Century Model*. Washington, DC: Science and Technology Innovation Program, Woodrow Wilson International Center for Scholars, April 2010.

Tucker, Jonathan, and Richard Danzig. *Innovation, Dual Use, and Security: Managing the Risks of Emerging Biological and Chemical Technologies*. Cambridge, MA: MIT Press, 2012.

Whiteside, Kerry. *Precautionary Politics: Principle and Practice in Confronting Environmental Risk*. Cambridge, MA: MIT Press, 2006.

Nanotechnology, virtual reality, synthetic biology

Allhof, Fritz, et al., eds. *Nanoethics: The Ethical and Social Implications of Nanotechnology*. Hoboken, NJ: Wiley, 2007.

Bainbridge, William Sims, and Mihail C. Roco, eds. *Managing Nano-Bio-Info-Cogno Innovations: Converging Technologies in Society*. New York: Springer, 2006.

Baldwin, Geoff, et al. *Synthetic Biology: A Primer*. London: Imperial College Press, 2012.

Bennett-Woods, Deb, ed. *Nanotechnology: Ethics and Society*. Boca Raton, FL: CRC Press, 2008.

Berube, David. *Nano-Hype: The Truth Behind the Nanotechnology Buzz*. Amherst, NY: Prometheus, 2006.

Blascovich, Jim, and Jeremy Bailenson. *Infinite Reality: The Hidden Blueprint of Our Virtual Lives*. New York: Morrow, 2011.

Carr, Nicholas. *The Shallows: What the Internet Is Doing to Our Brains*. New York: Norton, 2011.

Castronova, Edward. *Exodus to the Virtual World: How Online Fun Is Changing Reality*. New York: Palgrave-Macmillan, 2007.

Church, George, and Ed Regis. *Regenesis: How Synthetic Biology Will Reinvent Nature and Ourselves*. New York: Basic, 2012.

Drexler, K. Erik. *Engines of Creation: The Coming Era of Nanotechnology*. New York: Anchor, 1986, 1990.

ETC Group. *Extreme Genetic Engineering: An Introduction to Synthetic Biology*. Ottawa, Canada: ETC, January 2007.

McGonigal, Jane. *Reality Is Broken: Why Games Make Us Better and How They Can Change the World*. New York: Penguin, 2011.

Mulhall, Douglas. *Our Molecular Future: How Nanotechnology, Robotics, Genetics, and Artificial Intelligence Will Transform Our World*. Amherst, NY: Prometheus, 2002.

Roco, Mihail C., and William Sims Bainbridge, eds. *Converging Technologies for Improving Human Performance: Nanotechnology, Biotechnology, Information Technology, and Cognitive Science*. Dordrecht, Holland: Kluwer, 2003.

Schmidt, Markus, ed. *Synthetic Biology: Industrial and Environmental Applications*. Hoboken, NJ: Wiley-Blackwell, 2012.

Turkle, Sherry. *Alone Together: Why We Expect More from Technology and Less from Each Other*. New York: Basic, 2011.

Wohlsen, Marcus. *Biopunk: DIY Scientists Hack the Software of Life*. New York: Current, 2011.

Pharmaceutical enhancement

Barber, Charles. *Comfortably Numb: How Psychiatry Is Medicating a Nation*. New York: Pantheon, 2008.

Conrad, Peter. *The Medicalization of Society: On the Transformation of Human Conditions into Treatable Disorders*. Baltimore: Johns Hopkins University Press, 2007.

Elliott, Carl. *Better Than Well: American Medicine Meets the American Dream*. New York: Norton, 2003.

Gordijn, Bert, and Ruth Chadwick, eds. *Medical Enhancement and Posthumanity*. New York: Springer, 2008.

Greely, Henry, et al. "Towards Responsible Use of Cognitive-Enhancing Drugs by the Healthy." *Nature* 456 (2008): 702–5.

Kramer, Peter. *Listening to Prozac: A Psychiatrist Explores Antidepressant Drugs and the Remaking of the Self*. New York: Penguin, 1993.

Robots and artificial intelligence

Arkin, Ronald. *Governing Lethal Behavior in Autonomous Robots*. Boca Raton, FL: CRC Press, 2009.

Armstrong, George. *Smarter Than Us: The Rise of Machine Intelligence*. Berkeley, CA: MIRI, 2014.

Bar-Cohen, Yoseph, and Cynthia Breazeal, eds. *Biologically Inspired Intelligent Robots*. SPIE Publications, 2003.

Barrat, James. *Our Final Invention: Artificial Intelligence and the End of the Human Era*. New York: Thomas Dunne, 2013.

Bostrom, Nick. *Superintelligence: Paths, Dangers, Strategies*. New York: Oxford University Press, 2014.

Breazeal, Cynthia. *Designing Sociable Robots*. Cambridge, MA: MIT Press, 2002.

Brooks, Rodney A. *Flesh and Machines: How Robots Will Change Us*. New York: Pantheon, 2002.

Brynjolfsson, Erik, and Andrew McAfee. *The Second Machine Age: Work, Progress, and Prosperity in a Time of Brilliant Technologies*. New York: Norton, 2014.

Greenfield, Adam. *Everyware: The Dawning Age of Ubiquitous Computing.* Berkeley, CA: New Riders, 2006.

Hall, J. Storrs. *Beyond AI: Creating the Conscience of the Machine.* Amherst, NY: Prometheus, 2007.

Levy, David. *Love and Sex with Robots: The Evolution of Human-Robot Relationships.* New York: Harper, 2007.

Mazlish, Bruce. *The Fourth Discontinuity: The Co-Evolution of Humans and Machines.* New Haven, CT: Yale University Press, 1993.

Moravec, Hans. *Mind Children: The Future of Robot and Human Intelligence.* Cambridge, MA: Harvard University Press, 1988.

Rychlak, Joseph. *Artificial Intelligence and Human Reason: A Teleological Critique.* New York: Columbia University Press, 1991.

Singer, Peter. *Wired for War: The Robotics Revolution and Conflict in the 21st Century.* New York: Penguin, 2009.

Wallach, Wendell, and Colin Allen. *Moral Machines: Teaching Robots Right from Wrong.* New York: Oxford University Press, 2009.

Wolfe, Alan. *The Human Difference: Animals, Computers, and the Necessity of Social Science.* Oakland: University of California Press, 1993.

Science fiction

Anderson, M. T. *Feed.* Cambridge, MA: Candlewick Press, 2002.

Asimov, Isaac. *I, Robot.* New York: Bantam, 1991.

Atwood, Margaret. *Oryx and Crake.* New York: Doubleday, 2003.

Bly, Robert. *The Science in Science Fiction: 83 SF Predictions That Became Scientific Reality.* Dallas: BenBella, 2005.

Capek, Karel. *The Makropoulos Secret.* Boston: Luce, 1925.

Clarke, Arthur C. *2001: A Space Odyssey.* New York: Roc, 2000.

Clarke, Arthur C. *Childhood's End.* New York: Del Rey, 2001.

Crichton, Michael. *Prey.* New York: Avon, 2002.

Dick, Philip K. *Selected Stories of Philip K. Dick.* New York: Pantheon, 2002.

Gibson, William. *Neuromancer.* New York: Ace, 2000.

Halpern, James L. *The First Immortal.* New York: Del Rey, 1998.

Hanley, Richard. *Is Data Human? The Metaphysics of Star Trek.* New York: Basic, 1997.

Huxley, Aldous. *Brave New World.* New York: Harper, 1932.

Ishiguro, Kazuo. *Never Let Me Go.* New York: Knopf, 2005.

James, Edward, and Farah Mendlesohn, eds. *The Cambridge Companion to Science Fiction.* New York: Cambridge University Press, 2003.

Keyes, Daniel. *Flowers for Algernon.* Boston: Mariner, 2005 [first ed. 1959].

Kress, Nancy. *Beggars in Spain.* New York: Avon, 1993.

Orwell, George. *1984.* New York: Signet, 1949.

Powers, Richard. *Galatea 2.2.* New York: Harper, 1995.

Powers, Richard. *Generosity: An Enhancement.* New York: Farrar, Straus, and Giroux, 2009.

Shelley, Mary. *Frankenstein.* New York: Norton, 1995, 1818.

Stephenson, Neal. *Snow Crash.* New York: Bantam, 1992.

Science studies, history of technology, technological determinism

Allenby, Braden, and Daniel Sarewitz. *The Techno-Human Condition.* Cambridge, MA: MIT Press, 2011.

Bijker, Wiebe. *Of Bicycles, Bakelites, and Bulbs: Toward a Theory of Sociotechnical Change.* Cambridge, MA: MIT Press, 1997.

Clayton, Philip, and Paul Davies. *The Re-Emergence of Emergence: The Emergentist Hypothesis from Science to Religion.* New York: Oxford University Press, 2006.

Dupuy, Jean-Pierre. *On the Origins of Cognitive Science: The Mechanization of the Mind,* trans. M. B. DeBevoise. Cambridge, MA: MIT Press, 2009.

Gleick, James. *Faster: The Acceleration of Just About Everything.* New York: Vintage, 2000.

Haldane, J. B. S. "Daedalus, or, Science and the Future." Paper read to the Heretics, Cambridge, UK, February 4, 1923. http://vserver1.cscs.lsa.umich.edu/~crshalizi/Daedalus.html.

Hughes, Thomas. *Human-Built World: How to Think About Technology and Culture.* Chicago: University of Chicago Press, 2004.

Jasanoff, Sheila, et al., eds. *Handbook of Science and Technology Studies.* Thousand Oaks, CA: Sage, 1995.

Johnson, Deborah, and Jameson Wetmore, eds. *Technology and Society: Building Our Sociotechnical Future.* Cambridge, MA: MIT Press, 2009.

Kelly, Kevin. *What Technology Wants.* New York: Viking, 2010.

Kitcher, Philip. *Science in a Democratic Society.* Amherst, NY: Prometheus, 2011.

Latour, Bruno. *Nous n'avons jamais* été modernes [We Have Never Been Modern] Cambridge, MA: Harvard University Press, 1993.

Laughlin, Robert. *A Different Universe: Reinventing Physics from the Bottom Down.* New York: Basic, 2005.

Lessig, Lawrence. *Code: Version 2.0.* New York: Basic, 2006.

Lightman, Alan, Daniel Sarewitz, and Christine Desser, eds. *Living with the Genie: Essays on Technology and the Quest for Human Mastery.* Washington, DC: Island Press, 2003.

Mitchell, Melanie. *Complexity: A Guided Tour.* New York: Oxford University Press, 2009.

Moreno, Jonathan. *The Body Politic: The Battle over Science in America.* New York: Bellevue, 2011.

Nye, David. *Technology Matters: Questions to Live With.* Cambridge, MA: MIT Press, 2006.

Sewell, William. *Logics of History: Social Theory and Social Transformation.* Chicago: University of Chicago Press, 2005.

Smith, Merritt, and Leo Marx, eds. *Does Technology Drive History? The Dilemma of Technological Determinism.* Cambridge, MA: MIT Press, 1994.

Tenner, Edward. *Our Own Devices: How Technology Remakes Humanity.* New York: Vintage, 2003.

ACKNOWLEDGMENTS

I have tremendously enjoyed researching and writing this book over the past eleven years. It's a remarkable feeling now to look back and reflect on how many people have helped shape this project as it grew and shrank and morphed over time.

The research and writing were funded by a fellowship from the John Simon Guggenheim Memorial Foundation (2008–2009); a fellowship from the American Council of Learned Societies (2008–2009); a grant from the National Human Genome Research Institute, Program on Ethical, Legal, and Social Implications (Grant #RO3HG003298–01A1, 2005–2006); two grants from Vanderbilt's Research Scholar Grant Program (2008–2009, 2012–2013); a grant from Vanderbilt's Center for Ethics (2008); three sabbatical leaves from Vanderbilt's College of Arts and Science; and the endowment for the Chancellor's Chair in History at Vanderbilt. I could not have undertaken such a long project without the sustained financial support of these institutions, and I thank them most heartily.

Much of the pleasure I've derived from this project has come from the myriad conversations into which it has plunged me. The students in my Vanderbilt courses on human bioenhancement, Leonardo da Vinci, neuroscience, robotics, and human nature raised tough questions that pushed me to think harder. High school and middle school students to whom I gave presentations kindled my imagination with their probing comments. Colleagues and friends with whom I shared countless drafts of articles and chapters along the way gave unstintingly of their time and expertise. My thanks to you all!

Among the many persons whose influence marks these pages, I would like to single out the following for particularly warm thanks (needless to say, they are not responsible for the conclusions I reach): Susanna Barrows,

Andrea Baruchin, Tim Bent, Larry Churchill, Mark Cioc, Ellen Clayton, Jay Clayton, Beth Conklin, Joseph Davis, Jed Diamond, Marshall Eakin, M. Dominic Eggert, James Epstein, Sarah Farmer, Philippe Fauchet, Gerald Figal, Gary Gerstle, Jonathan Gilligan, Linda MacDonald Glenn, Michael Goldfarb, Richard Golsan, Vicki Greene, Gabrielle Hecht, David Holloway, Sarah Igo, Martin Jay, Carl Johnson, Fabrice Jotterrand, Peter Kuryla, Michel Laronde, Mary Lewis, Amanda Livsey, Billy Livsey, Kimberly Lomis, Elizabeth Lunbeck, W. Patrick McCray, Ole Molvig, Dan Morrison, Vivian Ota-Wang, Richard Alan Peters, Danielle Picard, Allison Pingree, Matthew Ramsey, Ruth Rogaski, Michael Rose, Anya Peterson Royce, J. B. Ruhl, Mark Scala, Jeff Schall, Vanessa Schwartz, Charles Scott, John Sloop, Alistair Sponsel, Helmut Smith, Michael Tangermann, Holly Tucker, Arleen Tuchman, Frank Wcislo, Meike Werner, Donald Worster.

I owe a special debt to my friends and colleagues who gave of their time to read the draft manuscript and offer me their comments on it: Nicholas Agar, Allen Buchanan, Joyce Chaplin, Kate Daniels, R. Samuel Deese, Tom Dillehay, Joel Harrington, Joseph Hinton, John Lachs, Geoff MacDonald, Vern Matheuszik, Keith Meador, Karen Misuraca, Simon Warwick-Smith, and David Wood. Their insights and suggestions made this a better book than it would have been.

Seven excellent research assistants helped me along the way, not only by gathering information, but also by debating key ideas with me and keeping me up to date on the latest scientific and engineering developments: Leslie Bruce, Maggie Corbett, Michael Greshko, Helen Li, Noah Fram, Jeff Berry, and Colleen Parker. I am grateful to Danielle Picard, Mike Petruna, Leslie Bruce, and Maggie Corbett for their terrific work on the book's companion wesite. A crucial role in the development of my ideas for this project was played by two yearlong faculty seminars at Vanderbilt University's Robert Penn Warren Center for the Humanities.

I have had the good fortune throughout my career to fall into the hands of gifted editors. Douglas Mitchell at the University of Chicago Press, and Ashbel Green at Knopf, taught me to be a better writer. My editor for this book, Helene Atwan of Beacon Press, is no exception: her myriad insightful observations, queries, and suggestions have improved the manuscript tremendously. Hearty thanks also to the entire team at Beacon Press for their splendid work in turning a manuscript into a book. My literary agent, Mildred Marmur, has sustained my writing through more than three decades with patience and good cheer; her expertise, encouragement, and humor, as well as her insider's knowledge of the publishing world, have been invaluable. My gratitude to my graduate school mentors at the University of California, Berkeley, Susanna Barrows, Martin Jay, and Michael Nagler, and to my meditation teachers, Joseph Goldstein, Jack Kornfield, and Stephen Levine, only grows deeper as the decades go by.

To my friends and extended family, who were forced for a decade to bear with my obsession with this material, I express my gratitude for their patience and for their love. Thank you, Western Besses, Daniels Clan, Italian Clan, Tracys, Allisons, McCrums. And thank you, old friends: Dave Baty, Hollis Cline, Jed Diamond, Virginia Griffith Frank, David Hurwith, Walden Kirsch, Doran Larson, Roberto Malinow, Bryan Tracy, Ruth Weizman.

As for you, Rina and Donovan, Kimberly, Natalie, and Sebastian:

Horizonless ocean
abyss of sky
Amid the trackless alien reaches of blind atom and chance
a dwelling emerges.

NOTES

Welcome to the Future

1. Wendell Berry, "The Sycamore," *The Selected Poems of Wendell Berry* (Berkeley, CA: Counterpoint, 1998), 27.

One: Envisioning the Future

1. Rainer Maria Rilke, *Letters to a Young Poet: Letter #8*, trans. M. D. Herter Norton (New York: Norton, 1934), 64–65.
2. See the online bibliography section on forecasting and futurology on this book's companion website, http://www.ourgrandchildrenredesigned.org.
3. Ernest Rutherford, quoted in Gertrud Weiss Szilard and Spencer Weart, *Leo Szilard: His Version of the Facts* (Cambridge, MA: MIT Press, 1978), 17.
4. I draw inspiration here from Fernand Braudel, *The Mediterranean and the Mediterranean World in the Age of Philip II*, trans. Sian Reynolds (1949 French ed.; New York: Harper & Row, 1972).
5. For discussions of prediction and forecasting specifically oriented toward technology, see the books by Seidensticker; Brockman; and Friedman in the online bibliography section on science studies.
6. The literary scholar Jay Clayton has observed that a significant subgenre of sci-fi novels written since WWII actually *do* explore the possibilities of a "posthuman" world in which humans have undergone myriad forms of modification—and that a majority of these novels actually consider such a scenario in a relatively positive light. I am confining my argument here to the novels and movies that have broken out successfully and influentially from this narrower audience. See Jay Clayton, "The Ridicule of Time: Science Fiction, Bioethics, and the Posthuman," *American Literary History* (March 2013): 1–27.
7. Technically, of course, the *Star Wars* stories do not take place in the future at all, but rather "a long time ago in a galaxy far, far away." But they nonetheless depict a technological world based on standard sci-fi extrapolations into the human future. George Lucas, *Star Wars* (Twentieth Century Fox, 1977–2005).
8. To be sure, plenty of excellent books within the sci-fi genre have explored issues of bioengineered populations of humanity, but they are not works that have become broadly influential in contemporary culture (not, at least, in the same sense as *Star Trek* or *Brave New World*). Among the most interesting recent examples of such

works are David Marusek, *Counting Heads* (New York: Tor, 2005); and the Hugo and Nebula award-winning novel by Nancy Kress, *Beggars in Spain* (New York: Avon, 1993). See Clayton, "The Ridicule of Time."

9. Ray Kurzweil, *The Age of Intelligent Machines* (Cambridge, MA: MIT Press, 1990); *The Age of Spiritual Machines: When Computers Exceed Human Intelligence* (New York: Penguin, 1999); *The Singularity Is Near: When Humans Transcend Biology* (New York: Viking, 2005); *How to Create a Mind: The Secret of Human Thought Revealed* (New York: Viking, 2012).

10. The term itself was first coined (in reference to its meaning as a unique historical turning point) in 1993 by a mathematician at San Diego State University, Vernor Vinge. In a speech before the Ohio Aerospace Institute, Vinge declared: "Within thirty years, we will have the technological means to create superhuman intelligence. Shortly after, the human era will be ended." For a text of the speech, see http://www-rohan.sdsu.edu/faculty/vinge/misc/singularity.html.

11. See Bill Joy, "Why the Future Doesn't Need Us," *Wired*, April 2000; and Kurzweil, *The Singularity Is Near*.

12. See the website of Humanity+, http://humanityplus.org/philosophy/philosophy-2/.

13. See the online bibliography section on enhancement and posthumanity on this book's companion website.

14. Kurzweil, *The Singularity Is Near*, chaps. 1–2.

15. Ibid., 15. This framework moves us into an entirely different scale of time, reminiscent of the grand visions of thinkers like G. W. F. Hegel, Henri Bergson, and Pierre Teilhard de Chardin: cosmic meta-narratives embedding human history within a broader process that unifies all phenomena, physical and immaterial, and that extends from the beginning of time into the distant future.

16. For a further discussion of Kurzweil, see appendix L on this book's companion website.

17. Bob Seidensticker, *Futurehype: The Myths of Technology Change* (San Francisco: Berrett-Koehler, 2006), 68–69.

18. See the books by Nye; Latour; Bijker, Hughes, and Pinch; Smith and Marx; and Jasanoff in the online bibliography section on science studies on this book's companion website.

19. A 2001 National Science Foundation study, the most recent publication of its kind, found that 5,580,200 people were employed in that year in the United States as scientists, engineers, and technicians (out of a total workforce of about 130 million). This amounts to about 4.2 percent of the total workforce. The specific numerical breakdown by field includes 524,800 managers, 2,157,300 scientists, 1,256,400 engineers, and 1,641,700 technicians. By comparison, the historical US employment data for these kinds of workers (with appropriate adjustments for differences in how professions were defined) yield the following rough statistics. Total workforce is given in parentheses, while the percentage of scientists, engineers, and technicians in the total workforce is given in brackets.

1850: 1,700 (5,277,000) [0.03 percent]; *1900*: 40,300 (27,554,100) [0.14 percent]; *1940*: 308,200 (47,584,200) [0.64 percent]; *1950*: 705,800 (63,870,600) [1.1 percent]; *1980*: 1,899,700 (97,378,400) [1.9 percent]; *2001*: 5,580,200 (130,000,000) [4.2 percent]

Source: National Science Foundation, Division of Science Resources, *Statistics. Scientists, Engineers, and Technicians in the United States: 2001,* NSF 05–313 (Arlington, VA: 2005); Bureau of Labor Statistics, US Department of Labor, *Occupational Employment Statistics,* table 2. Employment by Industry and Occupational Group, www.bls.gov/oes/; Matthew Sobek, "New Statistics on the U.S. Labor Force, 1850–1990," *Historical Methods* 34 (2001): 71–87, table A1.

Two: Pharmaceuticals

1. William Shakespeare, *Romeo and Juliet* (CreateSpace, 2014), act V, scene 3, 129.

2. For an overview of the scholarly debates over how to define this term, see appendix A online on this book's companion website, http://www.ourgrandchildrenredesigned.org.

3. The trend extends over the entire century, but has become especially pronounced since the 1980s. In 1980, private pharmaceutical research and development investments in the United States stood at $1.5 billion. By 1985, they had reached $6.8 billion. Since 1995, R&D spending has increased by at least $1 billion every year (some years produced multibillion-dollar increases). In 2005, spending reached $31 billion. In 1980, US pharmaceutical R&D stood at 4.85 percent of total private industry R&D. By 2006, the proportion reached nearly 15 percent. Drug sales totaled over $6.6 billion in 1970, $22.3 billion in 1980, $58.3 billion in 1990, $161 billion in 2000, and $245.8 billion in 2006. In twenty-two out of thirty-six years, sales grew at rates exceeding 10 percent; in thirty-one out of thirty-six years, sales growth exceeded 5 percent. American demand for pharmaceutical products is driving worldwide sales. From 1970 to 2006, the American market commanded an average 65.44 percent of total sales each year. Since 1960, the budget for the National Institutes of Health (which helps fund pharmaceutical research) has earned a steadily increasing proportion of total federal R&D spending. In 1960, the NIH received just 4 percent of all federal R&D spending; its 2007 budget accounted for over 29 percent. Mark P. Mathieu, ed., "R&D Spending by Research-Based Pharmaceutical Companies, 1980–2006," *Bio/Pharmaceutical R&D Statistical Sourcebook 2007/2008,* Parexel International Corporation (2008), 1; Mark Boroush and John E. Jankowski, eds., "National Patterns of R&D Resources, 2009 Data Update, NSF 12–321; Research and Development (R&D) Expenditures by Performing Sector," US National Science Foundation; NIH Almanac—Appropriations, http://www.nih.gov/about /almanac/appropriations/index.htm.

4. For example, take the case of tranquilizers. In the first half of the 1900s, a physician wishing to offer a patient a calming medication had to rely primarily on barbiturates, which were very effective at achieving sedation but tended to clobber the patient into a state of semi-stupor. A major breakthrough came in the 1940s with the drug meprobamate, marketed under the name Miltown. This newly synthesized drug achieved a much subtler soothing effect and rapidly became a best seller, earning more than $40 million a year for its manufacturer. But Miltown had a nasty side effect: it tended to be highly habit forming. As a result, a new class of tranquilizing drugs eventually came into being, the benzodiazepines, whose most famous protagonist was marketed as Valium. Though this drug, too, could prove habit forming, its soothing effects were more subtle, and its drawbacks far less pronounced, than any of its predecessors. By the early 1980s, therefore, it had become the single most

widely prescribed medication in the United States, with more than two billion doses sold. See the books by Elliott; Hoberman; Stein; and Tone in the online bibliography section on pharmaceutical enhancement on this book's companion website.

5. See, for example, Peter Conrad, *The Medicalization of Society: On the Transformation of Human Conditions into Treatable Disorders* (Baltimore: Johns Hopkins University Press, 2007).

6. Carl Elliott, *Better Than Well: American Medicine Meets the American Dream* (New York: Norton, 2003), 54–59.

7. See the excellent exposé of the aggressive promotional tactics used by pharmaceutical companies in Charles Barber, *Comfortably Numb: How Psychiatry Is Medicating a Nation* (New York: Pantheon, 2008), chaps. 2, 3.

8. John Hoberman, *Testosterone Dreams: Rejuvenation, Aphrodisia, Doping* (Oakland: University of California Press, 2005), chaps. 1–3; Micki McGee, *Self-Help, Inc.: Makeover Culture in American Life* (New York: Oxford University Press, 2005); Lennard Davis, *Enforcing Normalcy: Disability, Deafness, and the Body* (London: Verso, 1995).

9. The statistics, in this regard, speak for themselves. Cosmetic surgeries of various types increased fourfold in the United States during the 1990s (Elliott, *Better Than Well*, 189). They doubled in number between 1997 and 2011 ("15th Annual Cosmetic Surgery National Data Bank Statistics," American Society for Aesthetic Plastic Surgery, http://www.surgery.org/sites/default/files/ASAPS-2011-Stats.pdf). American physicians administered more than 1.6 million Botox injections in 2002. See Ramez Naam, *More Than Human: Embracing the Promise of Biological Enhancement* (New York: Broadway, 2005), 28. In 1999 the American pharmaceutical industry spent $13.9 billion on advertising and promotions; one company, Roche, spent $76 million promoting its weight-loss drug Xenical (Elliott, *Better Than Well*, 102, 119). The American Academy of Anti-Aging Medicine, founded in 1992, had grown to a membership of more than twenty thousand physicians and medical practitioners by 2009. (See the American Academy for Anti-Aging Medicine, http://www.worldhealth.net/pages /about.) During the early 2000s, Viagra sales generated $1 billion per year in revenue for its manufacturer, Pfizer. (Meika Loe, *The Rise of Viagra: How the Little Blue Pill Changed Sex in America* [New York: New York University Press, 2004], 8.) Americans spent $15 billion on dietary supplements in 2004. ("Dietary Supplements: A Framework for Evaluating Safety," Institute of Medicine of the National Academies, 2004, http://www.iom.edu/CMS/3788/4605/19578.aspx.)

10. Loe, *The Rise of Viagra*, 95–96.

11. Ibid., chap. 6.

12. Andy Coghlan, "First Anti-Wrinkle Pill Shows Signs of Success," *New Scientist*, September 21, 2012, http://www.newscientist.com/article/mg21128314.300-first -antiwrinkle-pill-shows-signs-of-success.html. See also the books by Blum; Gilman; and Etcoff in the online bibliography section on cosmetics on this book's companion website.

13. See the books by Jendrick; Assael; and Miah in the online bibliography section on cosmetics on this book's companion website.

14. See Naam, *More Than Human*, chap. 2; Julian Savulescu, Ruud ter Meulen, and Guy Kahane, eds., *Enhancing Human Capacities* (Hoboken, NJ: Wiley-Blackwell, 2011), chaps. 5–9; and the books by Harris; Buchanan; and Parens in the online

bibliography section on enhancement and posthumanity on this book's companion website. For a skeptical perspective, see Steven Rose, *The Future of the Brain: The Promise and Perils of Tomorrow's Neuroscience* (New York: Oxford University Press, 2005). See also Roy Jones et al., "Cognition Enhancers," Foresight Brain Science Project, British Government Office of Science and Technology, 2008; Gabriel Horn, "Brain Science, Addiction, and Drugs," British Academy of Medical Sciences, May 2008; Ross Anderson, "Why Cognitive Enhancement Is in Your Future (and Your Past)," *Atlantic*, February 6, 2012.

15. CREB stands for "cyclic adenosine monophosphate response element binding protein"—which presumably requires a significant exercise of memory formation merely to pronounce.

16. Ellen Gibson, "A Boom in Memory-Enhancing Drugs?" *Businessweek*, December 18, 2008; Melissa Healy, "Sharper Minds," *Los Angeles Times*, December 20, 2004. In 1990, biotech companies raised $900 million in private investment. They raised $4.2 billion in 1995, $6.7 billion in 1999, and $31.4 billion in 2000. All totaled, from 1980 to 2000, the industry raised over $249 billion in venture capital investment. "U.S. Venture Capital Disbursement, by Industry Category: 1980–2000," *Science and Engineering Indicators 2002*, National Science Foundation.

17. Michael Gazzaniga, *Human: The Science Behind What Makes Us Unique* (New York: Harper Collins, 2008); Neil Levy, *Neuroethics: Challenges for the 21st Century* (New York: Cambridge University Press, 2007), chaps. 4, 5.

18. What exactly do we mean when we speak of "cognitive enhancement"? Here it is useful to make a distinction among four complementary (and overlapping) aspects of human mental activity: *attention*: the ability to hold a representation or percept in our minds over extended periods, available for scrutiny or manipulation; *learning*: an ability to assimilate new information, concepts, and experiences; *memory*: the capacity to categorize, store, and retrieve what we have experienced or learned; and *insight*: creativity and innovation, the ability to make new connections, see things in novel ways, synthesize existing elements of knowledge into new wholes.

19. Andrew Jacobs, "The Adderall Advantage," *New York Times*, July 31, 2005.

20. Healy, "Sharper Minds."

21. Benedict Carey, "Brain Enhancement Is Wrong, Right?," *New York Times*, March 9, 2008.

22. Stephen Hall, "The Quest for a Smart Pill," *Scientific American*, September 2003, 7.

23. Healy, "Sharper Minds."

24. Aaron Cooper, "College Students Take ADHD Drugs for Better Grades," CNN. com, September 1, 2011; Judy Battista, "Drug of Focus Is at Center of Suspensions," *New York Times*, December 1, 2012. Another study, published in 2005, found that "the number of teenagers who admit to abusing prescription medications tripled from 1992 to 2003, while in the general population such abuse has doubled." Andrew Jacobs, "The Adderall Advantage," *New York Times*, July 31, 2005. Sales of modafinil in 2003 garnered more than $200 million for its manufacturer, Cephalon—a number that raised eyebrows because it suggested a far larger customer base than the total number of patients suffering from narcolepsy and related sleep disorders. The modafinil statistics, moreover, are dwarfed by those for Adderall and Ritalin: $4.7 billion in sales in 2007, with the rates of prescription

steeply on the rise. Ellen Gibson, "A Boom in Memory-Enhancing Drugs?" *Businessweek*, December 18, 2008.

25. Jonathan Haidt, *The Happiness Hypothesis: Finding Modern Truth in Ancient Wisdom* (New York: Basic, 2006), 39–40.

26. Leon Kass, *Beyond Therapy: Biotechnology and the Pursuit of Happiness* (New York: Regan, 2003), xvii.

27. Aldous Huxley, *Brave New World* (New York: Harper, 1932).

28. See the books by Stipp; Church; Medina; Rattan; Post and Binstock; Hall; De Grey; Arrison; and Zey in the online bibliography section on cosmetics, life extension, and physical enhancements on this book's companion website.

29. See appendix C on this book's companion website for a more detailed discussion of these "wear and tear" mechanisms.

30. Leonard Hayflick, "Modulating Aging, Longevity Determination, and the Diseases of Old Age," in *Modulating Aging and Longevity*, ed. Suresh Rattan (Dordrecht, Holland: Kluwer, 2003), chap. 1.

31. John Medina, *The Clock of Ages: Why We Age, How We Age, Winding Back the Clock* (New York: Cambridge University Press, 1996), chap. 14.

32. For a biographical profile of Hayflick, see Stephen Hall, *Merchants of Immortality: Chasing the Dream of Human Life Extension* (New York: Houghton Mifflin, 2003).

33. Andrea Bodnar et al., "Extension of Life-Span by Introduction of Telomerase into Normal Human Cells," *Science* 279, no. 5349 (January 16, 1998): 349.

34. Steven Austad, quoted in Naam, *More Than Human*, 84.

35. Piper Hunt et al., "Extension of Lifespan in *C. elegans* by Naphthoquinones That Act Through Stress Hormesis Mechanisms," *PLoS ONE* 6, no. 7 (July 13, 2011); Heidi Ledford, "Sirtuin Protein Linked to Longevity in Mammals: Male Mice Overproducing the Protein Sirtuin 6 Have an Extended Lifespan," *Nature*, February 22, 2012; Steven Post and Robert Binstock, eds., *The Fountain of Youth: Cultural, Scientific, and Ethical Perspectives on a Biomedical Goal* (New York: Oxford University Press, 2004), chaps. 7–11; Naam, *More Than Human*, chap. 4; Medina, *The Clock of Ages*, chap. 14.

36. Hayflick, "Modulating Aging," chap. 1.

37. Rose, quoted in Medina, *The Clock of Ages*, 312. See also Michael Rose, "The Metabiology of Life Extension," in *The Fountain of Youth*, chap. 7.

38. See appendix C, on this book's companion website, for a survey of research on radical life extension, and the online bibliography section on cosmetics, life extension, and physical enhancements.

39. For an overview of what scientists know today about the functional architecture of the human brain, see appendix B on this book's companion website.

Three: Bioelectronics

1. David Kellogg, *Inspector Gadget*, DVD (Disney, 1999, 2003).

2. Alison Abbott, "Mind-Controlled Robot Arms Show Promise," *Nature*, May 2012, http://www.nature.com/news/mind-controlled-robot-arms-show-promise-1.10652. For video, see "Paralyzed woman moves robot with her mind," Nature Video Channel (2012), http://www.youtube.com/watch?v=ogBX18maUiM&feature=player_embedded.

3. See the books by Katz; Nicolelis; Benford and Malartre; Chorost; Clark; and Smith and Morra in the online bibliography section on bioelectronics on this book's companion website, http://www.ourgrandchildrenredesigned.org.
4. I use the term more loosely and broadly in the present book than some other authors writing in this field do. See, for example, the books by Pethig and Smith; and Walker et al., in the online bibliography section on bioelectronics on this book's companion website.
5. Steven Kotler, "Vision Quest," *Wired*, September 2002.
6. William Dobelle, "Artificial Vision for the Blind by Connecting a Television Camera to the Visual Cortex," *American Society for Artificial Internal Organs [ASAIO] Journal* 46, no. 1 (February 2000): 3–9; ibid.
7. "Bionic Eye," CNN, June 2002. This video is not directly accessible through online search but has been posted on the website of the Brown University Artificial Retinas project. Using the taskbar on the left, click on "Cortical Implants," then "The Dobelle Implant."
8. Major labs currently working on restoration of sight include the Boston Retinal Implant Project (http://www.bostonretinalimplant.org) and Second Sight (http:// 2-sight.eu/en/home-en). Other notable studies on wireless implants have been pursued at Stanford University (Dr. James Loudin) and Tübingen University in Germany (Dr. Eberhart Zrenner). See http://www.nature.com/news/restoring-sight -with-wireless-implants-1.10627. See also an overview (as of 2006) of the major laboratories working on visual prostheses around the world at the Brown University Artificial Retinas project, http://biomed.brown.edu/Courses/BI108/2006–108websites /group03retinalimplants/index.htm. Still other researchers rely on the phenomena of synesthesia and brain plasticity, aiming to substitute signals from one sense for those of another. The idea is straightforward: take a pair of glasses that have a miniature camera mounted on them, and then process the visual signal stream through a computer, translating it into a series of electrical pulses. If you send the pulses to an auditory device worn over the ears, patients can learn the code for "seeing" representations of the outside world through specific patterns of sound. If you send the same pulses to a device resting on the tongue, the brain can also learn to read these coded patterns and deliver rudimentary spatial representations that correspond to visual scenes. Visualizing one's surroundings through one's ears or tongue: this approach does not, of course, deliver "sight" in the traditional sense of the word, but it has been adopted successfully by some blind persons as a practical tool for getting about and functioning more autonomously in the world. See the website of Seeing with Sound, http://www.seeingwithsound.com/. For the tongue-based synesthesia systems, see "Your Tongue Can See," ABC News, September 6, 2006, http://abcnews.go.com/Primetime/Story?id=2401551&page=1. See also Sunny Bains, "Mixed Feelings," *Wired*, March 2007, http://www.wired .com/wired/archive/15.04/esp_pr.html. See also James Geary, *The Body Electric: An Anatomy of the New Bionic Senses* (New Brunswick, NJ: Rutgers University Press, 2002).
9. Richard Normann, "Sight Restoration for Individuals with Profound Blindness," Center for Neural Interfaces, and John A. Moran Eye Center, University of Utah, http://www.bioen.utah.edu/cni/projects/blindness.htm.

10. "Meeting Focuses on Artificial Vision: 'The Eye and the Chip' Held in Detroit," *Retinal Physician*, August 2008; Jonathon Wells et al., "Transient Optical Nerve Stimulation: Concepts and Methodology of Pulsed Infrared Laser Stimulation of the Peripheral Nerve *In Vivo*," in *Neuroengineering*, ed. Daniel DiLorenzo and Joseph Bronzino (Boca Raton, FL: CRC Press, 2008), chap. 21, 1–19; Geary, *The Body Electric*.

11. Berlin Brain Computer Interface, http://www.youtube.com/watch?v=qCSSBEXBCbY.

12. Benjamin Blankertz et al., "The Berlin Brain-Computer Interface: Accurate Performance from First-Session in BCI-Naïve Subjects," *IEEE Transactions on Biomedical Engineering* 55, no. 10 (2008): 1–10; C. Guger et al., "How Many People Are Able to Operate an EEG-Based Brain-Computer Interface (BCI)?" *IEEE Transactions on Neural Systems and Rehabilitation Engineering* 11, no. 2 (June 2003): 145–47.

13. Timothy Hay, "Mind-Controlled Videogames Become Reality," *Wall Street Journal*, May 29, 2012.

14. Jonathan Wolpaw and Elizabeth Wolpaw, eds., *Brain-Computer Interfaces: Principles and Practice* (New York: Oxford University Press, 2012); Theodore Berger et al., *Brain-Computer Interfaces: An International Assessment of Research and Development Trends* (New York: Springer, 2008); Jan van Erp et al., "Brain-Computer Interfaces for Non-Medical Applications: How to Move Forward," *Computer* (IEEE Computer Society) 45, no. 4 (April 2012): 26–34, doi:10.1109/MC.2012.107; Guido Dornhege et al., eds., *Toward Brain-Computer Interfacing* (Cambridge, MA: MIT Press, 2007).

15. Debate rages in the bioengineering literature over which method of reading the brain's activities—invasive or noninvasive, implant or skullcap—will ultimately prove superior. Some scientists believe that the inherent limitations of externally situated skullcap interfaces will eventually present an insurmountable barrier to further improvements in the technology. They insist that truly advanced interactions between brains and machines will ultimately require a much higher bandwidth of communication into and out of the skull—and that this will favor the implanted technologies in the long run. Balderdash, say the proponents of skullcaps. They point to the tremendous increase in both the quality and quantity of the information they have proved capable of extracting from the brain over recent years, and argue that it is a mistake to devote scarce resources toward inherently dangerous implanted technologies. To which the implant proponents retort that further research will gradually render the invasive devices not only more effective, but also safer, cheaper, and more reliable as well. It is a classic technological debate, and the result is impossible to predict—except to say that *both* kinds of devices appear set to continue their headlong rush of advance over the coming decades. Wolpaw and Wolpaw, eds., *Brain-Computer Interfaces*; Berger et al., *Brain-Computer Interfaces*; Dornhege et al., eds., *Toward Brain-Computer Interfacing*.

     Another alternative to invasive and noninvasive brain monitoring technology lies in ECoG's (electrocortography). Devices are implanted invasively inside the skull but outside the brain (where the sensitive tissue lies). These devices yield significantly better signals than noninvasive skullcaps, and have a lower risk of compromising the brain's immunity than deep implants. For an overview, see Gerwin Schalk, "BCIs That Use Electrocorticographic Activity," in *Brain-Computer*

Interfaces, chap. 15; and Pagan Kennedy, "The Cyborg in Us All," *New York Times*, September 14, 2011.

16. Theodore Berger et al., "Brain-Implantable Biomimetic Electronics as a Neural Prosthesis for Hippocampal Memory Function," in *Toward Replacement Parts for the Brain: Implantable Biomimetic Electronics as Neural Prostheses*, ed. Theodore Berger and Dennis Glanzman (Cambridge, MA: MIT Press, 2005), 270.

17. Berger and Glanzman, eds., *Toward Replacement Parts for the Brain*.

18. Min-Chi Hsiao et al., "VLSI Implementation of a Nonlinear Neuronal Model: A "Neural Prosthesis" to Restore Hippocampal Trisynaptic Dynamics," *Proceedings of the 28th IEEE* (August 30–September 3, 2006): 4396–99. See also Theodore Berger et al., "Brain-Implantable Biomimetic Electronics for Restoration and Augmentation of Cognitive Function," IBM Almaden conference, PowerPoint, 2006, http://www.neural-prosthesis.com/index-9.html.

19. Theodore Berger, quoted in Eric Mankin, "Restoring Memory, Repairing Damaged Brains," USC News Service, June 17, 2011, http://news.usc.edu/#!/article/29100/Restoring-Memory-Repairing-Damaged-Brains.

20. Theodore Berger et al., "A Cortical Neural Prosthesis for Restoring and Enhancing Memory," *Journal of Neural Engineering* 8 (June 15, 2011), abstract.

21. Stephen Handelman, "The Memory Hacker," *Popular Science*, April 2007, 96.

22. Garrett Stanley et al., "Reconstruction of Natural Scenes from Ensemble Responses in the Lateral Geniculate Nucleus," *Journal of Neuroscience* 19 (September 15, 1999): 8036–42.

23. Yoichi Miyawaki et al., "Visual Image Reconstruction from Human Brain Activity using a Combination of Multiscale Local Image Decoders," *Neuron* 60 (December 11, 2008): 915–29.

24. The concept of "pleasure center" is no longer used by neuroscientists, since it has become clear over the intervening decades that (a) "pleasure" is too vague a term, and (b) there is no such thing as a unitary "pleasure center," because the experience of "pleasurable" sensations is modulated by a great many far-flung regions of the brain. Scientists now speak of "reward centers" to describe such sites. James Olds, "Self-stimulation of the Brain," *Science* 127, no. 3294 (February 14, 1958): 315–24.

25. Carl Sem-Jacobsen and Arne Torkildsen, "Depth Recording and Electrical Stimulation in the Human Brain," in *Electrical Studies on the Unanesthetized Brain*, ed. Estelle Ramey and Desmond O'Doherty (New York: Hoeber, 1960), 275–90; Jim Robbins, *A Symphony in the Brain: The Evolution of the New Brain Wave Feedback* (New York: Grove, 2008), 23.

26. Robert Heath, "Electrical Self-Stimulation of the Brain in Man," *American Journal of Psychiatry* 120 (December 1963): 571–77.

27. Charles Moan and Robert Heath, "Septal Stimulation for the Initiation of Heterosexual Behavior in a Homosexual Male," *Journal of Behavioral Therapy and Experimental Psychiatry* 3 (1972): 27. Heath then went on to conduct an experiment on B-19 that would be considered not only bizarre but extremely immoral by today's standards. B-19 was a twenty-four-year-old homosexual man who had been suffering from depression for many years. Heath decided he would try to kill two birds with one stone, "healing" B-19 of his homosexuality as well as his depression. B-19 was repeatedly shown pornographic videos depicting heterosexual sex while receiving

simultaneous electrical stimulation of his brain's reward centers. Over a series of weeks, Heath observed, B-19's overall mood and disposition seemed to become much cheerier and "more cooperative." At this point, Heath reports, "arrangements were made for a 21-year-old [female] prostitute to spend two hours with B-19 in a laboratory specially prepared to afford complete privacy." The results, as recounted in the *Journal of Behavioral Therapy and Experimental Psychiatry*, were everything that Heath had hoped: B-19 was able to have sex with a woman for the first time in his life, and subsequently "expressed how much he had enjoyed her and how he hoped that he would have sex with her again in the near future." The "treatment program" ended shortly afterward, and B-19 was released from the hospital. See Alan Baumeister, "The Tulane Electrical Brain Stimulation Program: A Historical Case Study in Medical Ethics," *Journal of the History of the Neurosciences* 9, no. 3 (2000): 262–78.

28. Robbins, *A Symphony in the Brain*, 25.

29. Helga Kuhse and Peter Singer, eds., *A Companion to Bioethics* (Malden, MA: Blackwell, 2001); Herbert Gottweis, *Governing Molecules: The Discursive Politics of Genetic Engineering in Europe and the United States* (Cambridge, MA: MIT Press, 1998). See also *From Birth to Death and Bench to Clinic: The Hastings Center Bioethics Briefing Book for Journalists, Policymakers, and Campaigns*, an excellent online resource on bioethics maintained by the Hastings Center, http://www.thehastingscenter.org/Publications/BriefingBook/Default.aspx.

30. Steven Marcus, ed., *Neuroethics: Mapping the Field* (New York: Dana, 2002); Judy Illes, ed., *Neuroethics: Defining the Issues in Theory, Practice, and Policy* (New York: Oxford University Press, 2006).

31. Kelvin Chou, Susan Grube, and Parag Patil, *Deep Brain Stimulation: A New Life for People with Parkinson's, Dystonia, and Essential Tremor* (New York: Demos Health, 2011); Adam Keiper, "The Age of Neuroelectronics," *New Atlantis* (Winter 2006): 18–19. This article also provides an excellent overview of bioelectronics in general.

32. Chou, Grube, and Patil, *Deep Brain Stimulation*; Bruce Katz, *Neuroengineering the Future: Virtual Minds and the Creation of Immortality* (Sudbury, MA: Infinity Science Press, 2008), 240; Wael Asaad and Emad Eskandar, "Deep Brain Stimulation for Obsessive Compulsive Disorder," in *Neuroengineering*, ed. DiLorenzo and Bronzino, chap. 9, 4.

33. One promising application of transcranial magnetic stimulation is in the treatment of depression. An Israeli company, Brainsway, is currently marketing a pioneering version of the technology. "Brainsway is counting on its Deep TMS System appealing to psychiatrists and patients because it offers a way to treat depression without surgery or pharmaceuticals that typically are taken for life. Regulators in the U.S. and Canada this month approved the product for use in patients who failed to respond to antidepressants or can't tolerate them." David Wainer, "Brainsway Sees Partnership as FDA Backs Depression Device," *Bloomberg Businessweek*, January 21, 2013.

34. A comprehensive recent survey is Carlo Miniussi, Walter Paulus, and Paolo Rossini, eds., *Transcranial Brain Stimulation* (Boca Raton, FL: CRC, 2012). See also Mark George and Edmund Higgins, *Brain Stimulation Therapies for the Clinician* (Washington, DC: American Psychiatric Publishing, 2008); Yiftach Roth and Abraham Zangen, "Transcranial Magnetic Stimulation of Deep Brain Regions," in

Neuroengineering, ed. DiLorenzo and Bronzino, chap. 22, 1–23; and Katz, *Neuro-engineering the Future*, 245–48.

35. Thomas Knopfel and Edward Boyden, eds., *Optogenetics: Tools for Controlling and Monitoring Neuronal Activity* (Amsterdam: Elsevier, 2012); Peter Hegemann and Stephan Sigrist, eds., *Optogenetics* (Boston: Walter de Gruyter, 2013).

36. David H. Freedman, "Brain Control," *MIT Technology Review*, October 27, 2010.

37. I am not referring here to electroshock therapy (EST) or to electro-convulsive therapy (ECT), which have been used since the 1930s by some psychiatrists to treat intractable depression. EST and ECT typically use an electrical current about four hundred times stronger than the stimulus applied in tDCS. They have not been used for cognitive enhancement purposes.

38. Miniussi et al., eds., *Transcranial Brain Stimulation*; Michael Nitsche et al., "Transcranial Direct Current Stimulation: State of the Art 2008," *Brain Stimulation* 1 (2008): 206–23; R. Douglas Fields, "Amping Up Brain Function: Transcranial Stimulation Shows Promise in Speeding Up Learning," *Scientific American*, November 25, 2011; Ross Andersen, "Why Cognitive Enhancement Is in Your Future (and Your Past)," *Atlantic*, February 6, 2012; Roi Kadosh et al., "The Neuroethics of Non-Invasive Brain Stimulation," *Current Biology* 22, no. 4 (February 21, 2012).

39. Fields, "Amping Up Brain Function."

40. Ibid.

41. Gerwin Schalk, "Brain-Computer Symbiosis," *Journal of Neural Engineering* 5 (2008): 1–15.

42. Timothy Hay, "Mind-Controlled Videogames Become Reality," *Wall Street Journal*, May 29, 2012.

43. Ian Rowley, "From Honda, A Mind-Reading Robot," *Businessweek*, March 31, 2009, http://www.businessweek.com/globalbiz/content/mar2009/gb20090331_865756 .htm?chan=top+news_top+news+index+-+temp_global+business.

44. The full quotation runs as follows: "Kennedy is often asked if his work will pave the way for real-world versions of Robocop. 'Even if that's where the science fiction is heading, I don't buy it,' he says. 'People think we can somehow stimulate patterns and generate thought, but we don't even know what a thought is yet. We know that neural activity is present when we're thinking and moving, but not much more.' He'll settle for giving a precious gift to the disabled: independence." Brendan Koerner, "Philip Kennedy: Melding man and machine to free the paralyzed," *US News and World Report*, January 3, 2000, 65.

45. Bioelectronic interface devices fall into two categories: open-loop versus closed-loop. Cathy Hutchinson's robotic arm prosthesis is an open-loop machine, because the electrical signals through which it operates only travel one way, from her brain to the machine. In order for this device to move into the closed-loop category, it would require sensors on the fingertips or elsewhere, sending signals back up the chain of connections, into the nervous system and up to the brain. This closed-loop system is how our own senses normally work, continually passing information to and fro between the brain and the sensory organs in an endless conversation, with the brain using this information to make ongoing adjustments in what it attends to and which follow-up instructions it sends back down the chain of command.

46. Sunny Bains, "Mixed Feelings," *Wired*, March 2007, http://www.wired.com/wired /archive/15.04/esp_pr.html.

47. Bijal P. Trivedi, "'Magnetic Map' Found to Guide Animal Migration," *National Geographic News*, October 12, 2001, http://news.nationalgeographic.com/news/2001/10 /1012_TVanimalnavigation.html.

48. See the books by Breazeal; Levy; Wallach and Allen; Arkin; and Kang in the online bibliography section on robotics and AI on this book's companion website.

49. As an analogy, I would point to the way in which my iPod has changed my relationship to music. I now possess a device on which every single song I have ever loved lies waiting for me. I can summon it up at any time. As a result, music has come to mean something quite different for me than it did a decade ago: it has been woven seamlessly into the texture of my daily life. I happen to think of a song, and ten seconds later I am listening to it. This changes my mood, which in turn leads me to behave differently than I would have. I am cheerier, or more wistful, or more pensive, than I would have been. The music jogs my memory, increasing my sense of connectedness to far-off moments of my life. I remember people, situations, predicaments, smells, places. As a result, the meaning of my passing experiences in the present seems more vivid, the colors in front of me more pronounced. The story of who I am feels more connected, more coherent. At one level, of course, the songs are still just songs, and I am still the same person as ever. But my whole relationship to music—and to myself—has subtly and profoundly shifted because of a simple, portable machine. I suspect that the knowledge-access-boosting skullcap would have a similarly potent effect.

50. Andy and Larry Wachowski, *The Matrix*, DVD (Warner, 1999). For the film script text, see http://www.scifiscripts.com/scripts/matrix_96_draft.txt.

51. James Cameron, *The Terminator*, DVD (MGM, 1984). For the film script text, see http://www.imsdb.com/scripts/Terminator.html.

52. See the books and articles in the online bibliography section on science fiction.

Four: Genetics and Epigenetics

1. Robert Sinsheimer, "The Prospect of Designed Genetic Change," *Engineering and Science* 32, no. 7 (April 1969): 11–12.

2. William K. Purves et al., *Life: The Science of Biology*, 5th ed. (Sunderland, MA: Freeman/ Sinauer, 1998), chaps. 10–12.

3. The definition I am giving here is a "quick and dirty" first approximation. Barry Barnes and John Dupré, in their excellent book *Genomes: And What to Make of Them* (Chicago: University of Chicago Press, 2008), 51–57, point out the many conceptual difficulties in defining what a gene really is. Matt Ridley, in *Nature via Nurture: Genes, Experience, and What Makes Us Human* (New York: Harper, 2003), 233–41, lays out seven distinct meanings of the term "gene."

4. Christopher Dickey, "I Love My Glow Bunny," *Wired*, April 2001.

5. Dan Charles, "Genetically Modified Corn Helps Common Kind, Too," National Public Radio, October 7, 2010, http://www.npr.org/templates/story/story.php?storyId =130405227; Philip Reilly, *Abraham Lincoln's DNA and Other Adventures in Genetics* (Cold Spring Harbor, NY: Cold Spring Harbor Laboratory Press, 2000), 159–60;

Graham Brookes and Peter Barfoot, *GM Crops: The First Ten Years: Global Socio-economic and Environmental Impacts* (Dorchester, UK: PG Economics, 2006).

6. Barnes and Dupré, *Genomes*; Eva Jablonka and Marion Lamb, *Evolution in Four Dimensions: Genetic, Epigenetic, Behavioral, and Symbolic Variation in the History of Life* (Cambridge, MA: MIT Press, 2005).

7. Barnes and Dupré, *Genomes*, 49–50.

8. See the books by Carey; Francis; Barnes and Dupré; and Jablonka and Lamb in the online bibliography section on genetics on this book's companion website, http://www.ourgrandchildrenredesigned.org.

9. Nessa Carey, *The Epigenetics Revolution: How Modern Biology Is Rewriting Our Understanding of Genetics, Disease, and Inheritance* (New York: Columbia University Press, 2012), 2.

10. Ibid.; Richard Francis, *Epigenetics: How Environment Shapes Our Genes* (New York: Norton, 2011).

11. Michael Rutter, *Genes and Behavior: Nature-Nurture Interplay Explained* (Malden, MA: Blackwell, 2006), especially chaps. 9, 10.

12. Erik Parens, Audrey Chapman, and Nancy Press, eds., *Wrestling with Behavioral Genetics: Science, Ethics, and Public Conversation* (Baltimore: Johns Hopkins University Press, 2006); Rutter, *Genes and Behavior*; Ridley, *Nature via Nurture*; and Judith Rich Harris, *No Two Alike: Human Nature and Human Individuality* (New York: Norton, 2006).

13. The notion of there being a single "gene for" such traits is rejected with vehemence and unanimity in the expert literature. See, for example, on this book's companion website, the books by Parens, Chapman, and Press; Rutter; Ridley; Pinker; and Harris in the online bibliography section on genetics.

14. Kenneth Schaffner, "Behavior: Its Nature and Nurture," in *Wrestling with Behavioral Genetics*, 10–11.

15. It is important to clarify here precisely what it means to "share" genes with another person. In the aftermath of the Human Genome Project, it has become common to find the following types of seemingly incommensurable assertions in discussions of genetics in the mass media:

> "Humans share 96 percent of their DNA with chimps, 70 percent of their DNA with the roundworm C. elegans, and 50 percent of their DNA with bananas." Stefan Lovgren, "Chimps, Humans 96 Percent the Same, Gene Study Finds," *National Geographic News*, August 31, 2005, http://news.nationalgeographic.com/news/2005/08/0831_050831_chimp_genes.html.

> "Any two humans randomly picked from around the world share 99.9 percent of their DNA." Neil Lamb, "The Human Genome Project: Looking Back, Looking Ahead," HudsonAlpha Institute for Biotechnology, Spring 2008, http://www.hudsonalpha.org/education/outreach/basics/hgp.

> "Human identical twins share 100 percent of their DNA, fraternal twins and ordinary siblings share 50 percent of their DNA, and adoptive siblings share none of their DNA." (See, for example, Harris, *No Two Alike*, 34–35: "Genetic similarity, defined as the proportion of genes shared by common descent, is 1.00 for identical twins, .50 for fraternal twins and ordinary siblings, and zero for adoptive siblings.")

The puzzled reader may conclude from these statements that I have more in common, genetically speaking, with the roundworm *C. elegans* than I do with my half-brother. The problem lies in imprecise terminology. In the comparison with other species, what researchers are referring to is the similarity in the total genomes after one lines up all the base pairs and compares them, side by side. In this sense of the word "sharing," humans possess many DNA sequences in common with other species because, like other species, our bodies are made up of cells, and a great deal of our genetic information is tailored to the basic developmental and biochemical tasks that are needed to run any multicellular organism. In the second sense of the word "sharing," researchers are referring to the specific (and relatively minuscule) subset of genetic alleles that vary between parents and their offspring, or between one sibling and another. This might be more accurately referred to as *familial* genetic variance; and in this sense of the word, two ordinary siblings can be said to share half their genes, because they each got a separately shuffled complement of twenty-three chromosomes from the same mother and father. This sets those siblings apart from their adoptive brother, all of whose forty-six chromosomes came from entirely different parents altogether.

16. Ridley, *Nature via Nurture*, 87–88, 90.
17. Ibid., 83. See also Schaffner, "Behavior: Its Nature and Nurture," in *Wrestling with Behavioral Genetics*, 10–11; Harris, *No Two Alike*, chap. 2.
18. Ridley, *Nature via Nurture*, 87–94.
19. Ibid., 83.
20. Eric Turkheimer et al., "Socioeconomic Status Modifies Heritability of IQ in Young Children," *Psychological Science* 14, no.6 (2003): 623–25.
21. See the books by Black; Kevles; Lombardo; Hawkins; and Carlson in the online bibliography section on genetics on this book's companion website.
22. Robert Proctor, *Racial Hygiene: Medicine Under the Nazis* (Cambridge, MA: Harvard University Press, 1988); Götz Aly, *Cleansing the Fatherland: Nazi Medicine and Racial Hygiene* (Baltimore: Johns Hopkins University Press, 1994).
23. Troy Duster, *Backdoor to Eugenics* (New York: Routledge, 2003); Nicholas Agar, *Liberal Eugenics: In Defence of Human Enhancement* (Malden, MA: Blackwell, 2004); Carlson, *The Unfit*, chap. 20; Black, *War Against the Weak*, chaps. 20–21; Daniel Kevles, *In the Name of Eugenics: Genetics and the Uses of Human Heredity* (Cambridge, MA: Harvard University Press, 2004), chaps. 17–19.
24. The most systematic articulation of this position is presented in Agar, *Liberal Eugenics*. See also the books by Naam; Stock; and Silver in the online bibliography section on enhancement and post-humanity.
25. Agar, *Liberal Eugenics*.
26. Reilly, *Abraham Lincoln's DNA and Other Adventures in Genetics*.
27. Ibid., 166–67.
28. Ibid., 180–81.
29. An excellent website on genetically modified organisms, sponsored by the European Union, is http://www.gmo-compass.org/eng/news/stories/.
30. See appendix D, on this book's companion website, for a survey of these pros and cons, and appendix M for a survey of contemporary genetic therapeutic techniques.

31. See the books by Nicholl; Wilmut; Javitt; Pence; and Harris in the online bibliography section on genetics on this book's companion website. See also the website on cloning maintained by the Human Genome Project at http://www.ornl.gov/sci /techresources/Human_Genome/elsi/cloning.shtml.

32. The clone's genome will not be completely identical to that of its donor organism, for two reasons. The somatic cell from which the clone was generated may contain genetic mutations that render it different from the germ cells of the organism from which it was taken. Moreover, the mitochondrial DNA in the cytoplasm of the denucleated egg will be different from the mitochondrial DNA in the cytoplasm of the donor's somatic cell. Scientists believe these two sources of genetic difference could play a significant role in determining the future success or failure of cloning efforts.

33. See the entry on cloning in Hastings Center, Bioethics Briefing Book, http://www .thehastingscenter.org/Publications/BriefingBook/Detail.aspx?id=2158.

34. See, on this book's companion website, the books by Harris; Macintosh; McGee; Mehlman; Nussbaum and Sunstein; Pence; and Wilmut in the online bibliography section on genetics.

35. Kerry Macintosh, *Human Cloning: Four Fallacies and Their Legal Consequences* (New York: Cambridge University Press, 2012); Troy Jollimore and Becky Cox White, "Multiplicity: A Study of Cloning and Personal Identity," in *Bioethics at the Movies*, ed. Sandra Shapsay (Baltimore: Johns Hopkins University Press, 2009), 102–20.

36. Eric Konigsberg, "Beloved Pets Everlasting?" *New York Times*, January 1, 2009.

37. Two fertility doctors, Severino Antinori and Panos Zavos, claimed, in 2001, to have successfully implanted cloned human embryos into women; however, since they refused to make public any data regarding their efforts, scientists have dismissed these reports as bogus. Emma Young and Damian Carrington, "Cloning Pregnancy Claim Prompts Outrage," *New Scientist*, April 5, 2002, http://www.newscientist.com/article /dn2133-cloning-pregnancy-claim-prompts-outrage.html. The following year, an organization called Clonaid, linked to the UFO-worshipping Raëlian cult, announced the birth of a healthy girl clone named Eve. Once again, no reliable evidence was forthcoming, and experts dismissed the claim as a publicity stunt. Gina Kolata and Kenneth Chang, "For Clonaid, a Trail of Unproven Claims," *New York Times*, January 1, 2003.

38. Ian Wilmut, *After Dolly: The Uses and Misuses of Human Cloning* (New York: Norton, 2006), chap. 7; Glenn McGee, *The Perfect Baby: Parenthood in the New World of Cloning and Genetics* (Lanham, MD: Rowman & Littlefield, 2000).

39. See the excellent worldwide data on public opinion and legislative matters surrounding biotechnology in general and cloning in particular, assembled by the Center for Genetics and Society, http://www.geneticsandsociety.org/article.php?id=401. See also Leon Kass, *Human Cloning and Human Dignity: The Report of the President's Council on Bioethics* (New York: PublicAffairs, 2002). For a strongly dissenting view, see Kerry Macintosh, *Illegal Beings: Human Clones and the Law* (New York: Cambridge University Press, 2005).

40. "About US Federal Policies & Human Biotechnology," Center for Genetics and Society, http://www.geneticsandsociety.org/article.php?list=type&type=44. See also the entry on cloning in Hastings Center, Bioethics Briefing Book.

41. Ya-Ping Tang et al., "Genetic enhancement of learning and memory in mice," *Nature* 401 (September 2, 1999): 63–69; Kristin Leutwyler, "Making Smart Mice," *Scientific American*, September 7, 1999, http://www.scientificamerican.com/article.cfm?id=making-smart-mice.

42. Ammar Hawasli et al., "Cyclin-dependent kinase 5 governs learning and synaptic plasticity via control of NMDAR degradation," *Nature Neuroscience* 10, no. 7 (July 2007), 880–86, http://ibg.colorado.edu/pdf/hawasli_et_al_07.pdf. See also Julie Steenhuysen, "Turning off gene makes mice smarter," Reuters, May 27, 2007, http://www.reuters.com/article/scienceNews/idUSN2546860920070527.

43. Consider, for example, the case of the Kaplan Test Prep company. This company was founded in 1938 by an enterprising man named Stanley Kaplan, who had been tutoring students in the basement of his Brooklyn home, helping them prepare for the formidable New York State Regents Exam. The company has gone through various vicissitudes since then, but the bottom line has been one of steady growth, with a particularly strong spurt during the 1990s. Revenues of Kaplan, Inc., for 2012 came to a whopping $2.2 billion. See http://www.washpostco.com/phoenix.zhtml?c=62487&p=irol-reportsannual.

   Kaplan's primary product is rather ethereal at one level, but fist-poundingly concrete at another. It offers the prospect of a competitive edge over other people taking the same test as you. Since test scores are based not only on your absolute performance, but also on your score relative to the scores of all the other test takers, students seeking admission to highly selective colleges and graduate schools rightly reach the following conclusion: if I can boost my score even as little as fifty points, compared to what I would have achieved without the test prep, my chances of getting into my dream school will be that much greater. Kaplan courses do not come cheap: the basic SAT preparation course costs $900 for twelve sessions, while the Rolls-Royce version with a private tutor will set you back $4,300. Yet large numbers of parents eagerly plunk down these kinds of sums every year: they want that edge for their kids.

44. For a discussion of possible techniques for altering the human germline in flexible and upgradeable ways, see appendix E on this book's companion website.

45. See the books in the online bibliography under the heading, "Genetics and Genetic Enhancement," especially those by Baillie and Casey; Buchanan; Burley and Harris; Chapman and Frankel; Cole-Turner; DeGrazia; Evans; Fletcher; Glannon; Gosden; Green; John Harris; Heyd; Macintosh; McGee; Mehlman; Nicholl; Nussbaum and Sunstein; Pence; Peterson; Shanks; Silver; and Stock.

46. Carey, *The Epigenetics Revolution*, chaps. 11–12; and Francis, *Epigenetics*, chap. 11.

47. Carey, *The Epigenetics Revolution*, chap. 11; and Francis, *Epigenetics*. See also the Human Epigenome Project, http://www.epigenome.org/index.php?page=consortium.

Five: Wild Cards

1. Stanley Kubrick, *2001: A Space Odyssey* (MGM, 1968).

2. Richard Feynman, "There's Plenty of Room at the Bottom," *Engineering and Science*, February 1960, 22–36, http://www.its.caltech.edu/~feynman/plenty.html; see also Chris Toumey, "Reading Feynman into Nanotechnology: A Text for a New Science," *Techné* 13, no. 3 (2008): 133–68.

3. K. Eric Drexler, *Engines of Creation: The Coming Era of Nanotechnology* (New York: Anchor, 1986, 1990).

4. See, on this book's companion website, http://www.ourgrandchildrenredesigned.org, the books by Drexler; Berube; Bainbridge and Roco; Foster; Hall; Milburn; Mulhall; Roco and Bainbridge; Sargent; Booker and Boysen; and Hodge, Bowman, and Maynard in the online bibliography section on nanotechnology, virtual reality, and synthetic biology.

5. Jerrod Kleike, ed., *National Nanotechnology Initiative: Assessment and Recommendations* (New York: Nova Science, 2010); John Sargent, *The National Nanotechnology Initiative: Overview, Reauthorization, and Appropriations Issues* (CreateSpace, 2012).

6. Earth's diameter: 12,742 kilometers. Glass marble's diameter: 1 or 2 centimeters. If you don't believe me, feel free to do the math.

7. See, on this book's companion website, the books by Berube; Foster; Hall; Milburn; Sargent; and Booker and Boysen in the bibliography section on nanotechnology.

8. Kewal Jain, *The Handbook of Nanomedicine* (Totowa, NJ: Humana, 2008); Robert Freitas, "Current Status of Nanomedicine and Medical Nanorobotics," *Journal of Computational and Theoretical Nanoscience* 2 (2005): 1–25; ETC Group, *Nanotech Rx: Medical Applications of Nano-scale Technologies: What Impact on Marginalized Communities?* (Ottawa, Canada: ETC, September 2006).

9. Jain, *The Handbook of Nanomedicine*.

10. James Baker and Istvan Majoros, eds., *Dendrimer-Based Nanomedicine* (Singapore: Pan Stanford, 2008), chap. 1; Kevin Bullis, "Nanomedicine," *Technology Review*, March/April 2006. See also Jain, *The Handbook of Nanomedicine*, 32.

11. Baker and Majoros, eds., *Dendrimer-Based Nanomedicine*, 6.

12. Drexler, *Engines of Creation*, 14, and passim.

13. See the excellent overview of these debates in David Berube, *Nano-Hype: The Truth Behind the Nanotechnology Buzz* (Amherst, NY: Prometheus, 2006), chap. 2.

14. Drexler, *Engines of Creation*, 172–73.

15. Kurt Vonnegut, *Cat's Cradle* (New York: Dell, 1998).

16. Berube, *Nano-Hype*, 55–57.

17. Paul Rincon, "Nanotech Guru Turns Back on 'Gray Goo,'" BBC News, June 6, 2004, http://news.bbc.co.uk/2/hi/science/nature/3788673.stm. See also Robert Freitas, "Some Limits to Global Ecophagy by Biovorous Nanoreplicators, with Public Policy Recommendations," Foresight Institute (April 2000), http://www.foresight.org/nano/Ecophagy.html.

18. See, on this book's companion website, the books by Allhof; Bennett-Woods; Sandler; and Hodge, Bowman, and Maynard in the online bibliography section on nanotechnology.

19. See the discussion in Deb Bennett-Woods, ed., *Nanotechnology: Ethics and Society* (Boca Raton, FL: CRC Press, 2008), chap. 8; and Fritz Allhof et al., eds., *Nanoethics: The Ethical and Social Implications of Nanotechnology* (Hoboken, NJ: Wiley, 2007), chaps. 7–14.

20. Bennett-Woods, ed., *Nanotechnology: Ethics and Society*; Allhof et al., eds., *Nanoethics*.

21. Rodolfo Llinás and Valeri Makarov, "Brain-Machine Interface via a Neurovascular Approach," in *Converging Technologies for Improving Human Performance:*

Nanotechnology, Biotechnology, Information Technology, and Cognitive Science, ed. Mihail Roco and William Bainbridge (Dordrecht, Holland: Kluwer, 2003), 245.

22. See, on this book's companion website, the books by Nilsson; Russell and Norvig; Pfeiffer and Scheier; Rychlak; Wolfe, Wallach, and Allen; McCorduck; Breazeal; Arkin; Kang; Brooks; Crevier; Moravec; and Minsky in the online bibliography section on robots and AI.

23. Rodney A. Brooks, *Flesh and Machines: How Robots Will Change Us* (New York: Pantheon, 2002); Cynthia Breazeal, *Designing Sociable Robots* (Cambridge, MA: MIT Press, 2002).

24. Brooks, *Flesh and Machines*, chaps. 4–6.

25. Nils Nilsson, *The Quest for Artificial Intelligence* (New York: Cambridge University Press, 2010); Stuart Russell and Peter Norvig, *Artificial Intelligence: A Modern Approach*, 3rd ed. (New York: Prentice Hall, 2010).

26. See the Roomba website, http://store.irobot.com/shop/index.jsp?categoryId =2804605.

27. In appendix P, on this book's companion website, I take up the question of whether it is wise to create entities that combine these kinds of powers with a capacity for intelligent, autonomous decision making and agency. See also the discussion in Nick Bostrom, *Superintelligence: Paths, Dangers, Strategies* (New York: Oxford University Press, 2014).

28. This is a much faster increase in efficacy, incidentally, than the rate predicted by Moore's Law in the realm of computer microprocessors. Robert Carlson, *Biology Is Technology: The Promise, Peril, and New Business of Engineering Life* (Cambridge, MA: Harvard University Press, 2010), 71–73; Erik Pettersson, Joakim Lundeberg, and Afshin Ahmadian, "Generations of Sequencing Technologies," *Genomics* 93 (2009): 105–11; see also "Sequencing Costs," National Human Genome Research Institute, http://www.genome.gov/sequencingcosts/.

29. Erik Brynjolfsson and Andrew McAfee, *The Second Machine Age: Work, Progress, and Prosperity in a Time of Brilliant Technologies* (New York: Norton, 2014), 179.

30. I discuss this phenomenon in Michael Bess, *The Light-Green Society: Ecology and Technological Modernity in France, 1960–2000* (Chicago: University of Chicago Press, 2003), chap. 2.

31. Frank Levy and Richard Murnane, *The New Division of Labor: How Computers Are Creating the Next Job Market* (Princeton, NJ: Princeton University Press, 2012).

32. Jeremy Rifkin, *The Zero Marginal Cost Society: The Internet of Things, the Collaborative Commons, and the Eclipse of Capitalism* (New York: Palgrave/Macmillan, 2014).

33. Brynjolfsson and McAfee, *The Second Machine Age*, chap. 12.

34. Arthur C. Clarke, interview by Gene Youngblood, *Los Angeles Free Press*, April 25, 1969, 42–43, 47; reprinted in *The Making of 2001: A Space Odyssey*, ed. Stephanie Schwam (New York: Modern Library, 2000), 258–69.

35. Brynjolfsson and McAfee, *The Second Machine Age*, chap. 14.

36. See, on this book's companion website, the books by Church and Regis; Baldwin; Bray; Carlson; ETC Group; Schmidt; Solomon; and Wohlsen in the online bibliography section on nanotechnology, virtual reality, and synthetic biology.

37. Drew Endy, quoted in Markus Schmidt and Camillo Meinhart, "Synbiosafe," documentary film (ISBN: 978–3-200–01623–1), (2009); see trailer, http://www.synbiosafe .eu/DVD/Synbiosafe.html.

38. Ibid.

39. Wil Hylton, "Craig Venter's Bugs Might Save the World," *New York Times*, May 30, 2012.

40. Geoff Baldwin et al., *Synthetic Biology: A Primer* (London: Imperial College Press, 2012); Robert Carlson, *Biology Is Technology: The Promise, Peril, and New Business of Engineering Life* (Cambridge, MA: Harvard University Press, 2010); ETC Group, *Extreme Genetic Engineering: An Introduction to Synthetic Biology* (Ottawa, Canada: ETC, January 2007); Markus Schmidt, ed., *Synthetic Biology: Industrial and Environmental Applications* (Hoboken, NJ: Wiley-Blackwell, 2012); Lewis Solomon, *Synthetic Biology: Science, Business, and Policy* (Piscataway, NJ: Transaction, 2012).

41. Marcus Wohlsen, *Biopunk: DIY Scientists Hack the Software of Life* (New York: Current, 2011).

42. An annual worldwide competition organized by MIT has emerged as the primary venue for these youths to showcase their insights and inventions. It is called iGEM (International Genetically Engineered Machine). The iGEM organization in Boston has established a registry of "standard biological parts" that anyone from Craig Venter to high school students can freely access online—a catalog of DNA sequences and other basic organic compounds and operators whose functional properties have been identified and tested. These parts are known as BioBrick components, and they serve as interchangeable elements that can be put together in myriad combinations to generate new cellular functions. In this way, Drew Endy's dream of "constructing organisms just like you construct bridges" is coming closer to reality. Baldwin et al., *Synthetic Biology*; Carlson, *Biology Is Technology*; Wohlsen, *Biopunk*. See the Parts Registry on the iGEM website, http://partsregistry.org/Main_Page.

43. All the books listed in note 40 above contain sections devoted to issues of safety, regulation, and control.

44. Drew Endy, quoted in ETC Group, *Extreme Genetic Engineering*, 23.

45. See the books cited above, particularly those by Church and Regis, Baldwin et al., Carlson, ETC Group, Schmidt, and Solomon.

46. Synthetic biologists convened a conference in Berkeley, California, in 2006, for which a key agenda item was addressing concerns about such dangers. They drafted a framework for self-governance among the field's practitioners, but in the end the conference's outcome remained inconclusive. See the overview of the draft policy statement at the following website, http://syntheticbiology.org/SB2.0/Biosecurity_ resolutions.html. For a thoughtful discussion of the pros and cons of regulating synthetic biology and other forms of cutting-edge biotechnology, see Carlson, *Biology Is Technology*, chap. 9. See also Richard Sklove, *Reinventing Technology Assessment: A 21st Century Model* (Washington, DC: Science and Technology Innovation Program, Woodrow Wilson International Center for Scholars, April 2010); Jonathan Tucker and Richard Danzig, *Innovation, Dual Use, and Security: Managing the Risks of Emerging Biological and Chemical Technologies* (Cambridge, MA: MIT Press, 2012).

47. William Bainbridge and Mihail C. Roco, eds., *Managing Nano-Bio-Info-Cogno Innovations: Converging Technologies in Society* (New York: Springer, 2006).

Six: Should We Reengineer the Human Condition?

1. Michel Foucault, "Truth, Power, Self: An Interview with Michel Foucault—October 25th, 1982," in *Technologies of the Self: A Seminar with Michel Foucault*, ed. L. H. Martin et al. (London: Tavistock, 1988), 9–15.

2. Stanley Hart and Alvin Spivak, *The Elephant in the Bedroom: Automobile Dependence & Denial: Impacts on the Economy and Environment* (Pasadena, CA: New Paradigm, 1993).

3. See the bibliography subheading of "Enhancement and post-humanity: bioethical and social implications."

4. Here are three representative articulations of the basic reasons why humans should not be enhanced:

> "The most significant threat posed by contemporary biotechnology is the possibility that it will alter human nature and thereby move us into a 'posthuman' stage of history. This is important because human nature exists, is a meaningful concept, and has provided a stable continuity to our experience as a species. It is, conjointly with religion, what defines our most basic values."
>
> —Francis Fukuyama, *Our Posthuman Future: Consequences of the Biotechnology Revolution* (New York: Farrar, Straus, and Giroux, 2002), 7.

> "We need to do an unlikely thing: we need to survey the world we now inhabit and proclaim it good. Good enough. Not in every detail; there are a thousand improvements, technological and cultural, that we can and should still make. But good enough in its outlines, in its essentials. We need to decide that we live, most of us in the West, long enough. We need to declare that, in the West, where few of us work ourselves to the bone, we have ease enough. In societies where most of us need storage lockers more than we need nanotech miracle boxes, we need to declare that we have enough stuff. Enough intelligence. Enough capability. Enough."
>
> —Bill McKibben, *Enough: Staying Human in an Engineered Age* (New York: Times Books, 2003), 109.

> "The problem with eugenics and genetic engineering is that they represent the one-sided triumph of willfulness over giftedness, of dominion over reverence, of molding over beholding. . . . The bigger stakes are of two kinds. One involves the fate of human goods embodied in important social practices—norms of unconditional love and an openness to the unbidden, in the case of parenting; the celebration of natural talents and gifts in athletic and artistic endeavors; humility in the face of privilege, and a willingness to share the fruits of good fortune through institutions of social solidarity. The other involves our orientation to the world that we inhabit, and the kind of freedom to which we aspire. It is tempting to think that bioengineering our children and ourselves for success in a competitive society is an exercise of freedom. But changing our nature to fit the world, rather than the other way around, is actually the deepest form of disempowerment."
>
> —Michael Sandel, *The Case Against Perfection: Ethics in the Age of Genetic Engineering* (Cambridge, MA: Harvard University Press, 2007), 85, 96–97.

5. Here are three representative quotations articulating the views of prominent pro-enhancement thinkers:

> "Only in mythical utopias or dystopias do we ever see a permanent sense of contentment. In the real world, contentment is transitory. It comes with each accomplishment. If it lasted too long, it would stunt our urge to grow and change. This hunger, this reach that exceeds our grasp, this aspiration to attain something 'which cannot be attained in earthly life' is the force that has built our world. It has built our comfortable lives. It has produced the medicine that keeps us alive. It has assembled our vast stores of knowledge. It has produced our art, our music, our philosophy. It has built our deepest understanding of the mysteries of the universe. Never to say enough, always to want more—that is what it means to be human."
>
> —Ramez Naam, *More Than Human: Embracing the Promise of Biological Enhancement* (New York: Broadway, 2005), 228.

> "I propose both the wisdom and the necessity of intervening in what has been called the natural lottery of life, to improve things by taking control of evolution and our future development to the point, and indeed beyond the point, where we humans will have changed, perhaps, into a new and certainly into a better species altogether. [. . . I defend] the idea of making people, or rather permitting people to make themselves and their children, longer-lived, stronger, happier, smarter, fairer (in the aesthetic and in the ethical sense of the term). [. . . I argue] not only for the freedom, but also for the obligation to pursue human enhancement."
>
> —John Harris, *Enhancing Evolution: The Ethical Case for Making Better People* (Princeton, NJ: Princeton University Press, 2007), 4–5, 9.

> "Transhumanists view human nature as a work-in-progress, a half-baked beginning that we can learn to remold in desirable ways. Current humanity need not be the endpoint of evolution. Transhumanists hope that by responsible use of science, technology, and other rational means we shall eventually manage to become post-human, beings with vastly greater capacities than present human beings have."
>
> —World Transhumanist Association, http://transhumanism.org/index.php/WTA/more/transhumanist-values/.

6. Among the countless books available, see Ernst Cassirer, *The Philosophy of the Enlightenment* (Princeton, NJ: Princeton University Press, 2009); and Peter Gay, *The Enlightenment* (New York: Norton, 1995).

7. Conor Cruise O'Brien, *The Great Melody: A Thematic Biography of Edmund Burke* (Chicago: University of Chicago Press, 1994); Thomas Sowell, *A Conflict of Visions: Ideological Origins of Political Struggles* (New York: Basic, 2007).

8. See appendix I on this book's companion website, http://www.ourgrandchildren redesigned.org, for a discussion of the scholarly debates concerning human nature and human personhood.

9. Sowell, *A Conflict of Visions*, chap. 2.

10. Ibid., 34.

11. Steven Pinker, *The Blank Slate: The Modern Denial of Human Nature* (New York: Viking, 2002), chap. 16.

12. Italics added by me. Alfred North Whitehead, *Process and Reality* (New York: Free Press, 1929, 1957), 400.

13. For a superb discussion of this topic, see Erik Parens, "Toward a More Fruitful Debate About Enhancement," in *Human Enhancement*, ed. Julian Savulescu and Nick Bostrom (New York: Oxford University Press, 2009), chap. 8.

14. Christopher Peterson and Martin Seligman, *Character Strengths and Virtues: A Handbook and Classification* (New York: Oxford University Press, 2004), chap. 1.

15. See, on this book's companion website, the books by Haidt; Csikszentmihalyi; Peterson; Peterson and Seligman; Kahneman, Diener, and Schwarz; and Lyubomirsky in the bibliography section on human nature, identity, and dignity.

16. Peterson and Seligman, *Character Strengths and Virtues*, chaps. 1–3.

17. Amartya Sen, *The Idea of Justice* (Cambridge, MA: Belknap, 2009); Christopher W. Morris, ed., *Amartya Sen* (New York: Cambridge University Press, 2009).

18. See Martha Nussbaum, *Creating Capabilities: The Human Development Approach* (Cambridge, MA: Harvard University Press, 2011); and the website of the United Nations Development Programme, http://hdr.undp.org/en/.

19. See appendix F, on this book's companion website, for a more detailed description of the rationale I followed in creating the list.

20. For a discussion of the concept of dignity, see appendix G, on this book's companion website.

21. Jonathan Haidt, *The Happiness Hypothesis: Finding Modern Truth in Ancient Wisdom* (New York: Basic, 2006).

22. See Larry Arnhart, *Darwinian Natural Right: The Biological Ethics of Human Nature* (Albany: SUNY Press, 1998).

Seven: Who Gets Enhanced?

1. Thomas Jefferson, letter to Roger C. Weightman, June 24, 1826, http://www.nobeliefs.com/jefferson.htm.

2. Jim Cullen, *The American Dream: A Short History of an Idea That Shaped a Nation* (New York: Oxford University Press, 2004); Michael Zweig, *The Working Class Majority: America's Best Kept Secret* (Ithaca, NY: ILR Press, 2001).

3. Narendra Jadhav, *Untouchables: My Family's Triumphant Escape from India's Caste System* (Oakland: University of California Press, 2007).

4. Jefferson, letter to Weightman.

5. Karl Drlica and David Perlin, *Antibiotic Resistance: Understanding and Responding to an Emerging Crisis* (Upper Saddle River, NJ: FT Press, 2011); Brad Spellberg, *Rising Plague: The Global Threat from Deadly Bacteria and Our Dwindling Arsenal to Fight Them* (New York: Prometheus, 2009).

6. David Landes, *The Unbound Prometheus: Technological Change and Industrial Development in Western Europe from 1750 to the Present*, 2nd ed. (New York: Cambridge University Press, 2003).

7. Ruth Cowan, *A Social History of American Technology* (New York: Oxford University Press, 1996); Sheila Jasanoff et al., eds., *Handbook of Science and Technology Studies* (Thousand Oaks, CA: Sage, 1995).

Eight: A Fragmenting Species?

1. Charles Darwin, *On the Origin of Species* (1859), http://darwin-online.org.uk /converted/pdf/1859_Origin_F373.pdf.

2. For a more detailed discussion of these issues, see appendix O on the companion website for this book, http://www.ourgrandchildrenredesigned.org.

3. "Popular Baby Names by Year," http://www.babycenter.com/babyNameYears .htm. See also Joyce Gramza, "Baby Names and Fads," May 4, 2009, http://www .sciencentral.com/video/2009/05/04/baby-names-and-fads/.

4. Historians and philosophers of science have made a compelling case in recent de- cades regarding the socially constructed meaning of biological traits in general and of genetically influenced traits in particular. One vivid example to which they often point is that of deafness. Most people tend to think of this as a straightforwardly bi- ological trait: either we can hear well, or not. If we cannot, then we have a handicap that we have to learn how to live with. It turns out, however, that things are not so simple. Many deaf persons ardently believe that their lack of hearing is not a handi- cap at all, but rather a form of social identity: deafness, for them, entails a different mode of constructing the sensory and social world, a mode possessing its own pro- found value that they would not want to give up for anything. Such persons actively choose to remain deaf, and to live within the rich culture of the deaf persons' com- munity, even when given the technological option of having their hearing restored. The biological ability to hear sounds, in other words, is not an absolutely desirable human trait. It may seem like such a trait to those of us who were born with func- tional ears and have lived our lives in the mainstream culture of the hearing. But it holds no such clear and unambiguous value to all persons reared within the distinc- tive culture of the deaf. Indeed, some deaf couples have attempted deliberately to engineer deafness into their offspring, in order to ensure that their children will be able to participate in the deaf community as full-fledged members. They fervently believe that they are *enhancing* their children by doing so. This issue continues to raise controversy, but regardless of the position we take on it, the broader conclu- sion we can draw here is that there is no such thing as a "purely biological" human trait. All our traits are simultaneously biological and cultural in nature, because we unavoidably interpret them for ourselves as we experience them and reflect on the roles they play in making us who we are. For books on the "social construction" of traits, see the works by Shilling; Hacking; Turner; Featherstone, Hepworth, and Turner; Barnes and Dupré; Fausto-Sterling; Kourany; Thacker; Sargent; DeGrazia; Bijker, Hughes, and Pinch; Locke and Farquhar in the online bibliography section on human nature, identity, and dignity, and disability studies. For works on deaf cul- ture, see Michael Parker, "The Best Possible Child," *Journal of Medical Ethics* 33, no. 5 (May 2007): 279–83; and, on this book's companion website, the books by Pad- den and Humphries; Lane; Davis; and Albrecht in the online bibliography section on human nature, identity, and dignity, and disability studies. See also Gregor Wolbring, "Confined To Your Legs," in *Living with the Genie: Essays on Technology and the Quest for Human Mastery*, ed. Alan Lightman et al. (Washington, DC: Island Press, 2003), chap. 8; and Gregor Wolbring, "Science and Technology and the Tri- ple D (Disease, Disability, Defect)," in *Converging Technologies for Improving Human Performance*, ed. Mihail C. Roco and William Sims Bainbridge (Dordrecht, Holland:

Kluwer, 2003), 232–43. This last article by Wolbring has an extensive bibliography on the social constructedness of disability.

5. Andres Duany, Elizabeth Plater-Zyberk, and Jeff Speck, *Suburban Nation: The Rise of Sprawl and the Decline of the American Dream* (New York: North Point Press, 2010).

6. Ibid., chaps. 3–6.

7. The philosopher Michel Foucault has contributed greatly to our understanding of these sorts of "homogenizing" or "normalizing" processes. His many works systematically explore the tacit norms embedded in each generation's most basic socialization processes; they examine the ways in which such norms come to inform, shape, and constrain the parameters of identity and morality that form the foundation for that era's social order. See Hubert Dreyfus and Paul Rabinow, *Michel Foucault: Beyond Structuralism and Hermeneutics* (Chicago: University of Chicago Press, 1983); and Michel Foucault, *Discipline & Punish: The Birth of the Prison*, trans. Alan Sheridan (New York: Vintage, 1995). See also Michael Bess, "Interview with Michel Foucault," https://my.vanderbilt.edu/michaelbess/foucault-interview/.

8. In contemplating what such a population would look like, it is also worth noting that there will probably be far fewer persons affected by birth defects, genetic anomalies, and inherited physical disabilities, because the advanced technologies that render bioenhancement possible will no doubt sharply reduce the incidence of such divergences from the physical norm. On the other hand, it would be unwise to assume that the enhancement technologies of the coming century will be completely free from failures of their own. I discuss the possibility for such "design errors" in chapter 9.

9. David Schneider, *The Psychology of Stereotyping* (New York: Guilford Press, 2005); Herbert Bless, Klaus Fiedler, and Fritz Strack, *Social Cognition: How Individuals Construct Social Reality* (East Sussex, UK: Psychology Press, 2004).

10. George Fredrickson, *Racism: A Short History* (Princeton, NJ: Princeton University Press, 2003); Michelle Alexander, *The New Jim Crow* (New York: New Press, 2012).

11. Helmut Smith, *The Continuities of German History: Nation, Religion, and Race Across the Long Nineteenth Century* (New York: Cambridge University Press, 2008), chaps. 4, 5; Neil Kressel, *Mass Hate: The Global Rise of Genocide and Terror* (New York: Westview, 2002); Fredrickson, *Racism*.

12. The Amish of today, however, enjoy one major advantage that the non-mods of tomorrow will probably not possess: they are united by a deep religious faith that is based on an ideology of love, acceptance, and humility. This faith not only helps structure the Amish community and hold it together over time; it also instills in Amish individuals the kinds of psychological resources that allow them to make peace with their humble and dependent position in American society. They do not aspire to have more than the deliberately simple life they have chosen for themselves and their children. Non-mods, by contrast, will have no such common ideological ground to rely on: their respective families will have become non-mods for a wide variety of reasons, many of them mutually incompatible. Although they will presumably be united in their fierce opposition to enhancement technologies, they will differ profoundly among themselves, and many of them will no doubt resent their structurally subordinate and marginal status within society. This will present a significant challenge to them, especially when it comes to perpetuating their community from one generation to the next: many of their children will no doubt reject the

non-mod choice of their elders, and sally forth into the enticing world of enhancements. The centrifugal forces operating within non-mod families and groups will be
considerably stronger than those among the Amish of today. John Hostetler, *Amish
Society* (Baltimore: Johns Hopkins University Press, 1993); Steven Nolt, *A History of
the Amish* (Intercourse, PA: Good Books, 2004).

13. Schneider, *The Psychology of Stereotyping;* Bless, Fiedler, and Strack, *Social Cognition.*
14. Misha Glenny, *The Fall of Yugoslavia: The Third Balkan War* (New York: Penguin,
    1996).

Nine: If I Ran the Zoo

1. Francisco de Goya y Lucientes, "The Sleep of Reason Produces Monsters," plate
   43, *The Caprices (Los Caprichos)* (18.64.43), in *Heilbrunn Timeline of Art History* (New
   York: Metropolitan Museum of Art, 2000), http://www.metmuseum.org/toah/works
   -of-art/18.64.43.
2. See, on this book's companion website, http://www.ourgrandchildrenredesigned.org,
   the books by Anthes; Bonnicksen; and Beauchamp and Frey (especially chaps. 11,
   12, 23, and 24) in the online bibliography section on animal enhancement.
3. R. Hightower, C. Baden, E. Penzes, P. Lund, and P. Dunsmuir, "Expression of Antifreeze Proteins in Transgenic Plants," *Plant Molecular Biology* 17 (1991): 1013–21.
4. Christopher Dickey, "I Love My Glow Bunny," *Wired*, April 2001.
5. Sanjiv Talwar et al., "Rat Navigation Guided by Remote Control," *Nature* 417 (May
   2, 2002): 37–38; Duncan Graham-Rowe, "Robo-Rat Controlled by Brain's Electrodes," *New Scientist*, May 1, 2002.
6. Bill Pohajdak et al., "Production of Transgenic Tilapia with Brockmann Bodies Se-
   creting [desThrB30] Human Insulin," *Transgenic Research* 13, no. 4 (2004): 313–23.
7. Henry Greely, "Human/Nonhuman Chimeras: Assessing the Issues," in *The Oxford
   Handbook of Animal Ethics*, ed. Tom L. Beauchamp and R. G. Frey (New York: Oxford University Press, 2011), 671–700; Maryann Mott, "Animal-Human Hybrids
   Spark Controversy," *National Geographic News*, January 25, 2005, http://news
   .nationalgeographic.com/news/2005/01/0125_050125_chimeras.html.
8. Will Ferguson, "Cyborg Tissue Is Half Living Cells, Half Electronics," *New Scientist*, August 28, 2012.
9. Technical definitions of the various animal/human genetic mixing possibilities are
   laid out by the bioethicist Julian Savulescu: "*Cybrids*: nuclear DNA is human,
   matched to the person from whom they were 'cloned'; very small amount of mitochondrial DNA from the animal's cytoplasm. *Transgenic embryos*: variable amount of
   animal DNA depending on how many genes are transferred. *Chimeras*: have both
   human and animal cells, because they are formed by merging human and animal
   embryos. *Hybrids*: have both human and animal chromosomes, because they are
   formed by fertilizing egg and sperm from different species." Julian Savulescu,
   "Genetically-Modified Animals: Should There Be Limits to Engineering the Animal
   Kingdom?," in *The Oxford Handbook of Animal Ethics*, 646. See also Andrea Bonnicksen, *Chimeras, Hybrids, and Interspecies Research: Politics and Policymaking* (Washington,
   DC: Georgetown University Press, 2009), 10–11; and Greely, "Human/Nonhuman
   Chimeras," in *The Oxford Handbook of Animal Ethics*, especially 673–78.

10. Gregory Stock, *Redesigning Humans: Choosing Our Genes, Changing Our Future* (Boston: Mariner, 2003), chaps. 3–6.

11. For an excellent discussion of such a scenario, see Greely, "Human/Nonhuman Chimeras," 671–700.

12. For a discussion of the boundary between humans and animals, see appendix J on this book's companion website. See also the books by Westoll; Hess; Wynne; Hauser; and Griffin in the online bibliography section on animal enhancement.

13. George Dvorsky, "All Together Now: Developmental and Ethical Considerations for Biologically Uplifting Nonhuman Animals," *Journal of Evolution and Technology* 18, no. 1 (May 2008): 129–42; James Hughes, *Citizen Cyborg: Why Democratic Societies Must Respond to the Redesigned Human of the Future* (New York: Westview, 2004), 221–27; David Langford, "Uplift," *The Greenwood Encyclopedia of Science Fiction and Fantasy* (Santa Barbara, CA: Greenwood, 2005), 850–51; Gorman Beauchamp, "The Island of Dr. Moreau as Theological Grotesque," *Papers on Language and Literature* 15 (1979): 408–17. See also the three "Uplift Saga" novels by David Brin: *Sundiver* (New York: Spectra, 1985); *Startide Rising* (New York: Spectra, 1984); and *The Uplift War* (New York: Spectra, 1987).

14. In Arthur C. Clarke's novel *2001: A Space Odyssey*, for example, the human species itself is revealed to have been the object of an Uplift intervention at the dawn of civilization: much of what is distinctively "sapiens" in *homo sapiens* can be traced to that seminal deed by our (still unknown) galactic benefactors. See Arthur C. Clarke, *2001: A Space Odyssey* (repr.; New York: Penguin Putnam, 2000).

15. From a certain perspective, the rearing of children in human society can be framed as an Uplift of sorts: we diligently steer the youngsters' development down certain socially acceptable paths, gradually elevating them to the level of fully functional adulthood. The key difference, of course, is that this act of elevation takes place among members of the same species and is framed within a model of development through stages (infancy, childhood, adulthood) *within* that species. The philosophy of Uplift commits the basic fallacy of applying this developmental concept *across different species*, as if they were each embodiments of distinct developmental stages within a broader, encompassing meta-species. But we have no evidence that such a meta-species really exists. The Uplifters, in short, fail to realize that their developmental metaphor is just a metaphor: they mistakenly believe that one species can embody the "infancy" of another, wholly distinct species.

16. The literature on imperialism and its critics is considerable. Some good places to start: Alice Conklin, *A Mission to Civilize: The Republican Idea of Empire in France and West Africa, 1895–1930* (Stanford, CA: Stanford University Press, 2000); Bernard Porter, *Critics of Empire: British Radicals and the Imperial Challenge* (New York: Tauris, 2008); A. P. Thornton, *The Imperial Idea and Its Enemies* (New York: Macmillan, 1985); Raymond Betts, *Assimilation and Association in French Colonial Theory, 1890–1914* (Lincoln: University of Nebraska Press, 2005).

17. Bonnicksen, *Chimeras, Hybrids, and Interspecies Research*; Savulescu, "Genetically-Modified Animals"; and Greely, "Human/Nonhuman Chimeras."

18. Sarah Chan and John Harris, "Human Animals and Nonhuman Persons," in *The Oxford Handbook of Animal Ethics*, ed. Tom Beauchamp and R. G. Frey (New York: Oxford

University Press, 2011), chap. 11; Michael Tooley, "Are Nonhuman Animals Persons?," in *The Oxford Handbook of Animal Ethics*, chap. 12.

19. See, on this book's companion website, the books by Grandin; Sunstein and Nussbaum; Oliver; and Beauchamp and Frey in the online bibliography section on animal enhancement.

20. Since you have taken the trouble to locate this endnote, you are obviously curious about vengefulness in a cat. Ashley lived in my college dorm. He belonged to my buddy Dave Baty. He was a huge creature with luxuriant light-gray fur—hence the moniker. If anyone mistreated Ashley or—to put it more accurately—if anyone behaved in a way that Ashley construed as mistreatment (for example, removing him from your bed before you settled in for the night), Ashley would not strike back immediately. He would bide his time. But sometime in the ensuing twenty-four hours, you could count on it. He would urinate in one of your shoes. Not anyone else's shoes. Just yours.

21. Donna Haraway, *Primate Visions: Gender, Race, and Nature in the World of Modern Science* (New York: Routledge, 1989); Donna Haraway, *When Species Meet* (Minneapolis: University of Minnesota Press, 2008); Cass Sunstein and Martha C. Nussbaum, eds., *Animal Rights: Current Debates and New Directions* (New York: Oxford University Press, 2004); Kelly Oliver, *Animal Lessons: How They Teach Us to Be Human* (New York: Columbia University Press, 2009); Beauchamp and Frey, *The Oxford Handbook of Animal Ethics*.

22. But now consider for a moment the exact opposite scenario. What if, instead of *boosting* the capabilities and sensibilities of animals, we were to drastically dumb them down? Given the often terrible conditions in which animals are raised and treated—whether on factory farms or in research laboratories—wouldn't it make sense, from a purely humanitarian point of view, to redesign such creatures so that they were incapable of suffering? This possibility is known in the bioethics literature as "disenhancement."

Suppose, for example, that you could generate a new form of chicken that just sat there all day in its tiny cage in a stupor, oblivious to its surroundings in the sprawling factory farm. Its brain and sensory system have been pared down through genetic engineering to the bare minimum required to sustain elementary life functions. It eats, it poops, it grows, but it otherwise possesses little or no discernible internal life. Once it is fully grown, it is killed and processed into meat for sale in your local grocery.

It is a prospect that most people today would find appalling; and for this reason, the idea of disenhancement is unlikely to get much traction in practice. Most voting citizens in modern societies have come to feel, at an intuitive level, that animals embody a kind of inherent value, and that this quality—while not the same as the value of humans—still renders them worthy of certain fundamental forms of respect. Bernard Rollin, *The Frankenstein Syndrome: Ethical and Social Issues in the Genetic Engineering of Animals* (New York: Cambridge University Press, 1995); Paul Thompson, "The Opposite of Human Enhancement: Nanotechnology and the Blind Chicken Problem," *Nanoethics* 2 (2008): 305–16; Arianna Ferrari, "Animal Disenhancement for Animal Welfare: The Apparent Philosophical Conundrums and the Real Exploitation of Animals. A Response to Thompson and Palmer," *Nanoethics* 6 (2012): 65–76; Emily

Anthes, *Frankenstein's Cat: Cuddling Up to Biotech's Brave New Beasts* (New York: Scientific American/Farrar, Straus, and Giroux, 2013).

23. Gary Albrecht et al., *Handbook of Disability Studies* (Thousand Oaks, CA: Sage, 2001); Martha Nussbaum, *Frontiers of Justice: Disability, Nationality, Species Membership* (Cambridge, MA: Belknap, 2006); Lennard Davis, *Enforcing Normalcy: Disability, Deafness, and the Body* (London: Verso, 1995); Lennard Davis, *My Sense of Silence: Memoirs of a Childhood With Deafness* (Champaign: University of Illinois Press, 2000); Jonathan Glover, *What Sort of People Should There Be?* (New York: Penguin, 1984).

24. See the discussion of international regulation in chapters 16 and 17. An instructive parallel challenge is that of restricting the misuse of biological and chemical weapons. See Jonathan Tucker and Richard Danzig, *Innovation, Dual Use, and Security: Managing the Risks of Emerging Biological and Chemical Technologies* (Cambridge, MA: MIT Press, 2012).

Ten: Mechanization of the Self

1. Lewis Thomas, "The Attic of the Brain," in *Late Night Thoughts on Listening to Mahler's Ninth Symphony* (New York: Viking, 1983), 141.

2. Peter Kramer, *Listening to Prozac: A Psychiatrist Explores Antidepressant Drugs and the Remaking of the Self* (New York: Penguin, 1993), 237–49.

3. Ibid., chap. 9.

4. See the insightful discussion of this factor in Carol Freedman, "Aspirin for the Mind? Some Ethical Worries About Psychopharmacology," in *Enhancing Human Traits: Ethical and Social Implications*, ed. Erik Parens (Washington, DC: Georgetown University Press, 1998), 135–50.

5. It is important to note here that Sonia's experience with Prozac was particularly positive and free of disturbing side effects. Many other individuals who took such psychotropic drugs found their symptoms worsened or came to feel as though they had become a completely different person (in a bad way). Some doctors and ethicists believe that these kinds of medications are being overprescribed in contemporary society, and that we should completely reassess the way we use them. These are important considerations, and they are being hotly debated by patients, psychiatrists, and bioethicists today. See Charles Barber, *Comfortably Numb: How Psychiatry Is Medicating a Nation* (New York: Pantheon, 2008); Joseph Glenmullen, *Prozac Backlash: Overcoming the Dangers of Prozac, Zoloft, Paxil, and Other Antidepressants with Safe, Effective Alternatives* (New York: Simon & Schuster, 2001); Peter Breggin, *The Anti-Depressant Fact Book: What Your Doctor Won't Tell You About Prozac, Zoloft, Paxil, Celexa, and Luvox* (New York: Da Capo, 2001).

6. Joseph Forgas, Roy Baumeister, and Dianne Tice, eds., *Psychology of Self-Regulation: Cognitive, Affective, and Motivational Processes* (New York: Psychology Press, 2009); Daniel Wegner, *White Bears and Other Unwanted Thoughts: Suppression, Obsession, and the Psychology of Mental Control* (New York: Guilford, 1994); Ana Guinote and Theresa Vescio, eds., *The Social Psychology of Power* (New York: Guilford, 2010).

7. Quoted in Sherry Turkle, "Sociable Technologies: Enhancing Human Performance When the Computer is not a Tool but a Companion," in *Converging Technologies for Improving Human Performance: Nanotechnology, Biotechnology, Information Technology, and Cognitive Science*, ed. Mihail Roco and William Sims Bainbridge (Dordrecht,

Holland: Kluwer, 2003), 150. See also Sherry Turkle, *Alone Together* (New York: Basic, 2011), 10.

8. See, on this book's companion website, http://www.ourgrandchildrenredesigned.org, the discussion of human nature as compared with "animal nature" in appendix J, online.

9. Neil Levy, *Neuroethics: Challenges for the 21st Century* (New York: Cambridge University Press, 2007), 78.

10. A possible source of confusion here, Peter Kramer suggests, lies in the concept itself of "mood brightener." A mood-brightening drug, as most people understand the term, would be a chemical that allows me, at 1:00 p.m., to take a pill because I'm feeling blue, and to find myself, at 1:10 p.m., feeling just peachy. But Prozac, it turns out, does not work this way at all. It does not deliver a specific state of affect on demand, as if one were putting on rose-tinted glasses. According to Kramer, when Prozac worked well for one of his patients (which was far from being always and reliably the case), it often functioned more as a globally transformative agent operating on the individual's entire personality. "Prozac often surprises us," he noted. "Sometimes it will change only one trait in the person under treatment; but often it goes far beyond a single intended effect. You take it to treat a symptom, and it transforms your sense of self." Kramer, *Listening to Prozac*, 267–68.

11. Judith Hooper and Dick Teresi, *The Three-Pound Universe* (New York: Tarcher/Putnam, 1986), 155–56.

12. Charlie Chaplin, *Modern Times* (United Artists, 1936).

13. Robin Leidner, *Fast Food, Fast Talk: Service Work and the Routinization of Everyday Life* (Oakland: University of California Press, 1993); Arlie Hochschild, *The Managed Heart: Commercialization of Human Feeling*, 2nd ed. (Oakland: University of California Press, 2012); Ludwig Von Mises, *Bureaucracy* (Important Books, 2012); Michael Lipsky, *Street-Level Bureaucracy: Dilemmas of the Individual in Public Service* (New York: Russell Sage, 2010); Michael Volkin, *The Ultimate Basic Training Guidebook: Tips, Tricks, and Tactics for Surviving Boot Camp*, 2nd ed. (El Dorado Hills, CA: Savas Beatie, 2009).

Eleven: Turbocharging Moral Character

1. Kurt Vonnegut, *Slaughterhouse Five* (New York: Dell, 1969), 154.

2. Peter Singer and Agata Sagan, "Are We Ready for a 'Morality Pill'?" *New York Times*, January 28, 2012, http://opinionator.blogs.nytimes.com/2012/01/28/are-we-ready-for-a-morality-pill/.

3. Paul Zak, "The Trust Molecule," *Wall Street Journal*, April 27, 2012, http://online.wsj.com/article/SB10001424052702304811304577365782995320366.html; Paul Zak, *The Moral Molecule: The Source of Love and Prosperity* (New York: Dutton, 2012); Michael Kosfeld, Markus Heinrichs, Paul Zak, Urs Fischbacher, and Ernst Fehr, "Oxytocin Increases Trust in Humans," *Nature* 435 (June 2, 2005): 673–76.

4. Zak, *The Moral Molecule*, chaps. 1–2.

5. Ibid., 46.

6. Not all researchers have reached the same optimistic conclusions about oxytocin as Zak, however. See Carsten De Dreu, Lindred Greer, Gerben Van Kleef, Shaul Shalvi, and Michel Handgraaf, "Oxytocin Promotes Human Ethnocentrism," *Proceedings of the National Academy of Sciences* 108, no. 4 (January 25, 2011): 1262–66.

7. For a discussion of the requirements for meaningful moral action, see appendix H on this book's companion website, http://www.ourgrandchildrenredesigned.org. See also the discussion in Ingmar Persson and Julian Savulescu, *Unfit for the Future: The Need for Moral Enhancement* (New York: Oxford University Press, 2012); John Harris, "Moral Enhancement and Freedom," *Bioethics* 25, no. 2 (2011): 102–11; and Lenny Moss, "Moral Molecules, Modern Selves, and Our 'Inner Tribe,'" *Hedgehog Review* 15, no. 1 (Spring 2013): 19–33.

8. Persson and Savulescu, *Unfit for the Future*, 111–14.

9. See, on this book's companion website, the books by Copp; LaFollette; MacIntyre; Nuccetelli and Seay; Rachels; Rawls and Kelly; Sandel; Scanlon; Schaler; Scheffler; Sen; Singer; Wiggins; and Wilson in the online bibliography section on justice and ethics.

10. In Scenario 2 (unenhanced altruism), by contrast, your decision to act involves a genuine choice among many morally distinctive options. You assess the situation, rapidly weighing the risks facing you and the moral imperatives incumbent on you; images of fleeing to safety flash through your mind, as well as images of rushing into danger to help the old lady; you experience a variety of emotions; you come to a decision and spring into action. The full range of options, from egoistic to altruistic, has become real and plausible for you, and you have ultimately opted for a good deed. This internal decision process certainly includes the release of a variety of brain chemicals, but those neurochemical reactions arise through the activity of your own higher faculties, such as reason, empathy, moral judgment, and the weighing of risks and consequences. Your brain processes are not skewed in one direction (empathy) by the addition of an extraneous chemical: rather, when empathy arises in your mind, it confronts you as one among many impulses and feelings competing for primacy in the field of your attention. Your dispositional states therefore reflect more faithfully the mutual interaction of multiple, self-generated mental processes arising within the ongoing whole that is your personhood. In this sense, your action has a quality of authenticity, of moral genuineness, that is missing in Scenario 3.

11. John Harris, "Moral Enhancement and Freedom," *Bioethics* 25, no. 2 (2011): 102–11; Robert Kane, ed., *The Oxford Handbook of Free Will*, 2nd ed. (New York: Oxford University Press, 2011).

12. The only true enhancement of our moral action is one that augments our mental resources for exercising free will and autonomy—for making our own reflective choices about how to behave, based on serious consideration of the fullest practical range of options. In an alternative rendering, therefore, we might envision the morality pill quite differently, as a chemical for boosting our *faculties for making autonomous choices*. To the extent that the pill broadened the mental space in which we envision possibilities for ourselves, to the extent that it deepened the range of factors we could reflectively take into account—such a pill would be truly enriching the moral character of our agency. It would augment precisely those features of our sovereign decision-making processes that render them most precious and meaningful to us, and that distinguish us from animals and machines. Of course, the behavior of a person who used this latter kind of choice-empowering pill would still retain the fundamental quality that necessarily characterizes all moral agents: the nature of her actions would be unforeseeable, uncertain. In the end, the decisions made by such an enhanced individual might well prove wiser and more beneficent that those of a

nonmodified one. But, then again, they might not. They would remain, in their essence, unpredictable. That is one of the hallmarks of genuine free will, when it is exercised in the context of fully functional human personhood.

Twelve: Shared Intimacies

1. Rajesh P. N. Rao and Andrea Stocco, "When Two Brains Connect: The Dawn of Brain-to-Brain Communication Has Arrived," *Scientific American Mind*, November-December 2014, 36.

2. See, on this book's companion website, http://www.ourgrandchildrenredesigned.org, the books by Rose; Dehaene; DiLorenzo and Bronzino; Katz; Moreno; and Raey-maekers in the online bibliography section on brain/mind/neuroethics. See also appendix B online.

3. Douglas Heaven, "Telepathic Animals Solve Task as One," *New Scientist*, March 9, 2013, 8–9; Miguel Pais-Vieira, Mikhail Lebedev, Carolina Kunicki, Jing Wang, and Miguel Nicolelis, "A Brain-to-Brain Interface for Real-Time Sharing of Sensorimotor Information," *Scientific Reports* 3, no. 1319 (February 28, 2013): 1–10, http://www.nature.com/srep/2013/130228/srep01319/full/srep01319.html; Miguel Nicolelis, *Beyond Boundaries: The New Neuroscience of Connecting Brains with Machines—and How It Will Change Our Lives* (New York: Times Books, 2011).

4. Seung-Schik Yoo et al. "Non-invasive Brain-to-Brain Interface," *PLOS One*, April 3, 2013, http://www.plosone.org/article/info:doi/10.1371/journal.pone.0060410.

5. Rao and Stocco, "When Two Brains Connect," 36–39.

6. Ibid., 38–39.

7. One recent example is *Paycheck*, a movie about a man who works on top-secret engineering projects and has his memory selectively erased each time a project is finished, so that he cannot divulge the nature of the devices he has created. The story depicts the memory-erasing process as a straightforward scrolling through a linear sequence of richly detailed scenes, each scene encapsulating one second of the man's conscious awareness that is stored in his brain. The erasure then consists of simply deleting a mnemonic time segment, just as if one were erasing a section of a videotape. For all our vast ignorance about how memory works, we can say one thing with near certainty: this depiction of memory's functioning is balderdash. John Woo, *Paycheck*, DVD (Paramount, 2003).

8. See, on this book's companion website, the books by Kandel; Tononi; Bear; Koch; Damasio; Edelman; LeDoux; Frith; Gazzaniga; and Hooper and Teresi in the online bibliography section on brain/mind/neuroethics.

9. Some neuroscientists believe that this multifaceted and wide-ranging associative mechanism may underlie the basic processes through which abstract categories are formed in the mind. Joe Z. Tsien, "The Memory Code: Learning to Read Minds by Understanding How Brains Store Experiences," *Scientific American*, July 2007, 52–59.

10. Andy Clark, *Natural-Born Cyborgs: Minds, Technologies, and the Future of Human Intelligence* (New York: Oxford University Press, 2003), chap. 5.

11. To be sure, one could also argue that such a sharing of memories would merely reveal, in starker contrast and greater detail, the deep differences that *divide* individuals and cultures from each other. My basic premise, however, is that the differences between people are more easily apparent to each other than the similarities: people

have no problem identifying all the things that render the Other distinct, weird, alien, separate. It is generally much harder for them to recognize the underlying commonalities that link them with other groups at a deeper level. The memory-sharing technology would probably allow this underappreciated realm of overlapping and similar features to become more salient in people's experience of each other.

12. David DeGrazia, *Human Identity and Bioethics* (New York: Cambridge University Press, 2005); John Perry, ed., *Personal Identity* (Oakland: University of California Press, 1975); Charles Taylor, *Sources of the Self: The Making of the Modern Identity* (Cambridge, MA: Harvard University Press, 1992); Antonio Damasio, *Self Comes to Mind: Constructing the Conscious Brain* (New York: Pantheon, 2010); Derek Parfit, *On What Matters*, 2 vols. (New York: Oxford University Press, 2011).

13. The robot designer Hans Moravec acknowledges, and celebrates, this likely outcome in his two visionary works: *Mind Children: The Future of Robot and Human Intelligence* (Cambridge, MA: Harvard University Press, 1988); and *Robot: Mere Machine to Transcendent Mind* (New York: Oxford University Press, 1999).

14. Timothy Trull and Mitch Prinstein, *Clinical Psychology* (Independence, KY: Wadsworth, 2012); Andrew Pomerantz, *Clinical Psychology: Science, Practice, and Culture* (Los Angeles: Sage, 2010).

15. Pierre Teilhard de Chardin, *The Phenomenon of Man* (New York: Harper, 1959).

Thirteen: Virtual Reality and (Yawn) This Other Reality

1. Edward Castronova, *Exodus to the Virtual World: How Online Fun Is Changing Reality* (New York: Palgrave-Macmillan, 2007), 200–201.

2. Jane McGonigal, *Reality Is Broken: Why Games Make Us Better and How They Can Change the World* (New York: Penguin, 2011), 3. See also, on this book's companion website, http://www.ourgrandchildrenredesigned.org, the books by Castronova; Boellstorff; Nardi; and Taylor in the online bibliography section on nanotechnology, virtual reality, and synthetic biology.

3. Edward Castronova, *Synthetic Worlds: The Business and Culture of Online Games* (Chicago: University of Chicago Press, 2005), 75.

4. Jim Blascovich and Jeremy Bailenson, *Infinite Reality: The Hidden Blueprint of our Virtual Lives* (New York: Morrow, 2011), chaps. 2–4.

5. McGonigal, *Reality Is Broken*, chaps. 1–2; Blascovich and Bailenson, *Infinite Reality*, chaps. 2–4.

6. The gaming industry in 2012 boasted revenues of about $68 billion—a figure greater than the GDP of 140 member-nations of the UN. McGonigal, *Reality Is Broken*, 4.

7. This is a common theme in the works by McGonigal, Castronova, Boellstorff, Nardi, and Taylor.

8. I will employ the terms "real life" and "primary reality" to refer to the physically grounded world into which each of us is born. The idea itself of "primary reality" is of course problematic. From Plato to Kant to Freud to the deconstructionists and cognitive researchers of the present day, philosophers and psychologists have struggled with this elusive concept. I will not attempt here to engage that still-ongoing discussion, which cuts across epistemology, metaphysics, and ontology. Instead, I will "bracket" this issue, following the lead of pragmatically oriented philosophers

who adopt a stance of critical and provisional realism. Most humans operate in their daily lives on the assumption that there is an independent, external reality, whose causal processes and observable regularities set the parameters for their body's survival as a biological organism, and for the myriad arrangements of their social existence. At the same time, most thoughtful persons eventually come to recognize that this external reality—as experienced by humans—is also very much a malleable construct: its features are profoundly influenced by the perspective of the person who is observing it, and by that person's particular mental faculties or cultural dispositions. Threading a path between these two primordial poles of the objective and the subjective, most of us make our daily way. The literature on these matters is, of course, gargantuan. A good place to start: Paul Moser, ed., *The Oxford Handbook of Epistemology* (New York: Oxford University Press, 2005); Michael Loux and Dean Zimmerman, eds., *The Oxford Handbook of Metaphysics* (New York: Oxford University Press, 2005); David Chalmers, David Manley, and Ryan Wasserman, eds., *Metametaphysics: New Essays on the Foundations of Ontology* (New York: Oxford University Press, 2009).

9. As neuroscience and bioelectronics continue their rapid advance, moreover, these two fields will no doubt exert a significant influence on virtual reality technologies. One profound possibility, in this regard, would be for researchers to bypass the external senses altogether and switch to directly stimulating the various sensory centers in the brain itself. Instead of surrounding our bodies with immersive simulation devices, we would then simply connect ourselves to a brain-machine interface, allowing the software to instill the experiencing of specific scenarios directly into our minds. This is, of course, a purely speculative idea at present; but if such devices did exist, they would amount to what the philosopher Robert Nozick once famously described as an "experience machine." Robert Nozick, *Anarchy, State, and Utopia* (New York: Basic, 1977), 42–45. Such an apparatus would presumably deliver a simulation so perfect that it would be completely indistinguishable from real-life experiences. Steven Rose, *The Future of the Brain: The Promise and Perils of Tomorrow's Neuroscience* (New York: Oxford University Press, 2005), chaps. 10–11; Mihail Roco and William Bainbridge, eds., *Converging Technologies for Improving Human Performance: Nanotechnology, Biotechnology, Information Technology, and Cognitive Science* (Dordrecht, Holland: Kluwer, 2003); William Bainbridge and Mihail Roco, eds., *Managing Nano-Bio-Info-Cogno Innovations: Converging Technologies in Society* (New York: Springer, 2006); Bruce Katz, *Neuroengineering the Future: Virtual Minds and the Creation of Immortality* (Sudbury, MA: Infinity Science Press, 2008).

10. Telepresence, robotics, and haptic feedback technologies, which today allow a surgeon to operate on a patient lying in a hospital on the other side of the world, would further augment the effect of simulation devices. Seeing and hearing would then be complemented by a wide variety of tactile sensations, further heightening the realism of the experience. When you smote your enemy in the virtual world, you would feel a corresponding physical sensation in your hand and arm, transmitted back through the interface device to your body. William Sherman and Alan Craig, *Understanding Virtual Reality: Interface, Application, and Design* (San Francisco: Morgan Kaufmann, 2003).

11. Oliver Bimber and Ramesh Raskar, *Spatial Augmented Reality: Merging Real and Virtual Worlds* (Wellesley, MA: Peters, 2005); Rolf Hainrich, *The End of Hardware:*

Augmented Reality and Beyond, 3rd ed. (Booksurge, 2009); James Kent, *The Augmented Reality Handbook* (Tebbo, 2011).

12. In 2012, the Google corporation made a splash with a video it released, previewing what it would feel like to wear its new Google Glass, an interactive headpiece worn like spectacles. See the Google Glass website, https://plus.google.com/+projectglass #+projectglass/posts; for the video, http://www.youtube.com/watch?v=JSnBo6um5r4. In a similar vein, two MIT researchers recently unveiled a new wearable device, dubbed SixthSense, that acts as a "portable personal assistant." It consists of a small camera, a pocket projector, a mirror, and a mobile Web-linked computer, all of which can be worn as a single unit, pendant-style, around the neck. Operating the interface is simple: you wear color-coded caps on your thumb and index finger on each hand, and the computer picks up your hand motions through the camera. Make a little box-shaped gesture with your fingers, framing a portion of the scene in front of you, and the machine snaps a photo of that part of the view. Browse through a newspaper, and the computer uses character recognition to read the headline being displayed, and then supplements the print version of the article with the latest updates from the Web, projecting images and text down onto the page as you hold it. Pick up a book in a bookstore, and the machine projects reviews and ratings right onto the page. The device creates what amounts to a personalized "information-space" surrounding the user wherever she goes. Pranav Mistry, "SixthSense: Integrating information with the real world," Fluid Interfaces Group, MIT Media Lab (2009), http://www.pranavmistry.com/projects/sixthsense /index.htm#PUBLICATIONS.

For a video laying out the capabilities of the device in detail, see "Pattie Maes & Pranav Mistry: Unveiling the 'Sixth Sense,' Game-changing Wearable Tech," http:// www.ted.com/index.php/talks/pattie_maes_demos_the_sixth_sense.html.

13. A remarkable technology currently under development would incorporate some of these features into a contact lens. Babak Parviz, a researcher in electrical engineering at the University of Washington, announced in 2008 that he and his team had successfully tested on rabbits a "bionic contact lens" that incorporates ultra-thin light-emitting diodes, embedded electronic circuitry, and a small antenna. The prototype is not yet fully functional, but the goal, according to Parviz, is to develop a device that projects a computer interface directly into a person's field of vision, allowing individuals to manipulate objects on virtual screens suspended in midair. Parviz has now gone on to work for Google on the Google Glass project. See Bryn Nelson, "The Vision of the Future Seen in Bionic Contact Lens," MSNBC.com, January 21, 2008, http://www.msnbc.msn.com/id/22731631/. The "virtual murals" I described above already adorn the buildings of today. Using a smart phone app called Aurasma, you can overlay images, artwork, text, or video onto any physical object you wish, including other people. You take your phone's camera and frame a particular location you wish to adorn—say, a restaurant where you just encountered the surliest serving staff you've ever met. As you walk away, you tag the image and daub the restaurant with your negative rating and any other comments you wish to leave; the GPS on the phone marks the site and records your feedback. Any future user of Aurasma who walks by that restaurant will see your comments pop up on her iPhone. Put this app together with Google Glass, and you have a restaurant that

shouts out at people as they walk past, "This place sucks, go away!" (Needless to say, such technologies will pose new challenges to property law and the concept of free speech.) See Catherine de Lange, "Reality Re-Spray: The Hidden Digital World Uncloaked," *New Scientist*, May 25, 2012.

14. I calculated this figure by multiplying the estimated number of gamers worldwide (660 million) by the average number of their weekly online hours (thirteen hours), extended over a one-year period. See the statistics in McGonigal, *Reality Is Broken*, "Introduction."

15. Ibid., chap. 3.

16. Ibid., "Introduction."

17. Employers, for example, could devise systems for providing continual constructive feedback to their workers, along with structured advancements through ascending levels of mastery; teachers could design their students' assignments more along the lines of collaborative quests; roommates could reinvent the daily household chores as a contest operating under arcane and amusing rules. The common goal: to take some of the most alluring features of gaming and reinject them into those aspects of primary reality that tend to be fraught with meaningless toil and drudgery. See Mc-Gonigal, *Reality Is Broken*, part III. In 2009, the *Guardian* applied this concept of "gaming in the real world" to stunning effect. After allegations surfaced about wide-spread abuse of expense accounts by members of parliament, Her Majesty's Government released a massive trove of digitized expense documents for the journalists to peruse. There was a catch, however: the documents were completely unorganized, and there were 458,832 of them. Each receipt or expense record would have to be scrutinized, one by one. There was no way the *Guardian*'s beleaguered staff could carry out such a task, but instead of giving up, they turned to a well-known game developer in London, who told them quite simply, "Make it feel like a game." (McGonigal, *Reality Is Broken*, 222.) Once the task had been moved online and revamped as an entertaining "Catch your MP with his pants down" contest, more than twenty thousand persons eagerly participated, frantically scouring the records for malfeasance. The end result: a bevy of resignations and criminal proceedings, followed by reform of official accounting procedures. As a cherry on top, the nabbed MPs were obliged to repay the government 1.1 million pounds. Score one for the gamers—in primary reality.

18. Blascovich and Bailenson, *Infinite Reality*, 102–8.

19. Ibid.

20. Timothy Judge and Daniel Cable, "The Effect of Physical Height on Workplace Success and Income: Preliminary Test of a Theoretical Model," *Journal of Applied Psychology* 89, no. 3 (June 2004): 428–41.

21. See the discussion in Kevin Kelly, *What Technology Wants* (New York: Viking, 2010).

22. For purposes of brevity, I am condensing and melding here some of the many sophisticated arguments put forth by Sherry Turkle in *Alone Together: Why We Expect More From Technology and Less From Each Other* (New York: Basic, 2011), and Jaron Lanier, *You Are Not a Gadget: A Manifesto* (New York: Knopf, 2010).

23. Turkle, *Alone Together*, 295.

24. See, in particular, Lanier, *You Are Not a Gadget*, chap. 10.

25. Castronova, *Exodus to the Virtual World*, chaps. 9–10.

26. The literature on this subject, not surprisingly, is vast. Some good places to start: Uri Merry, *Coping with Uncertainty: Insights from the New Sciences of Chaos, Self-Organization, and Complexity* (New York: Praeger, 1995); Donald Norman, *Living With Complexity* (Cambridge, MA: MIT Press, 2010); Philip Hodgson and Randall White, *Relax, It's Only Uncertainty: Lead the Way When the Way Is Changing* (London: Pearson, 2001); David Wilkinson, *The Ambiguity Advantage: What Great Leaders Are Great At* (New York: Palgrave-Macmillan, 2006).

Fourteen: Till Death Do Us No Longer Part

1. Joel Coen and Ethan Coen, *The Big Lebowski* (Working Title Films), 1998.
2. See, on this book's companion website, http://www.ourgrandchildrenredesigned.org, the books by Stipp; Medina; Overall; Rattan; Post and Binstock; Hall; De Grey; Arrison; Zey; Maher and Mercer; and Immortality Institute in the online bibliography section on cosmetics, life extension, and physical enhancements.
3. For a description of these strategies, see appendix C online. Suresh Rattan, ed., *Modulating Aging and Longevity* (Dordrecht, Holland: Kluwer, 2003); Michael Rose, "The Metabiology of Life Extension," in *The Fountain of Youth: Cultural, Scientific, and Ethical Perspectives on a Biomedical Goal*, ed. Steven Post and Robert Binstock (New York: Oxford University Press, 2004), chap. 7; Robert Arking, Extending Human Longevity: A Biological Probability," in *The Fountain of Youth*, chap. 8; Richard Miller, "Extending Life: Scientific Prospects," in *The Fountain of Youth*, chap. 10; Aubrey De Grey, *Ending Aging: The Rejuvenation Breakthroughs That Could Reverse Human Aging in Our Lifetime* (New York: St. Martin's, 2007).
4. Some promoters of anti-aging research like to point out the fact that the average life expectancy of humans has *already doubled* in recent times: it hovered around an age of thirty-one years as recently as the year 1850. However, this statistic is misleading: the increase in average life expectancy since that time has been almost entirely due to a reduction in infant and childhood mortality, *not* to a major increase in how long people lived after having survived through childhood. Once you made it to age twenty in Victorian England, you were likely to live into your sixties or even your seventies if you took good care of yourself. The advance of science and medicine since that time has only exerted a relatively modest effect on life expectancy after the onset of adulthood. See Thomson Prentice, "Health, History, and Hard Choices: Funding Dilemmas in a Fast-Changing World," World Health Organization, August 2006, www.who.int/global_health_histories/seminars/presentation07.pdf.
5. Steven Post and Robert Binstock, "Introduction," in *The Fountain of Youth*, 2–3.
6. See, for example, Michael Rose, "The Metabiology of Life Extension," in *The Fountain of Youth*, chap. 7 (Rose is a professor of evolutionary biology at the University of California, Irvine); Robert Arking, "Extending Human Longevity: A Biological Probability," in *The Fountain of Youth*, chap. 8 (Arking is a professor of biology and gerontology at Wayne State University); Richard Miller, "Extending Life: Scientific Prospects," in *The Fountain of Youth*, chap. 10 (Miller is a professor of pathology and associate director of the Geriatrics Center at the University of Michigan); and John Medina, *The Clock of Ages: Why We Age; How We Age; Winding Back the Clock* (New York: Cambridge University Press, 1996). Medina is a molecular biologist at the University of Washington.

7. Joel Cohen, *How Many People Can the Earth Support?* (New York: Norton, 1995); Ian Angus and Simon Butler, *Too Many People? Population, Immigration, and the Environmental Crisis* (Chicago: Haymarket, 2011); Charles Kenny, *Getting Better: Why Global Development Is Succeeding—and How We Can Improve the World Even More* (New York: Basic, 2011).

8. Sonia Arrison, *100 Plus: How the Coming Age of Longevity Will Change Everything, from Careers and Relationships to Family and Faith* (New York: Basic, 2011), 53.

9. In Sweden, for example, the fertility rate today is 1.8 children per woman: the nation's total population has been slowly declining over recent decades. However, if all Swedes were suddenly given an "immortality pill" that prevented them from aging beyond healthy adulthood, and these immortals kept having babies at the current rates, the nation's total population would increase by only 22 percent by the year 2100. The low fertility rate significantly diminishes the impact of the vastly lower death rate (some "immortals" would presumably still be dying from accidents, suicides, and incurable diseases). Gavrilov and Gavrilova explain: "Common sense and intuition would predict a demographic catastrophe if immortal people were to continue to reproduce; that is what we initially believed too. However, a deeper mathematical analysis leads to paradoxical results. If parents produce less than two children on average, so that each next generation is smaller by some common ratio (R < 1), then even if everybody were immortal, the size of the population over time would not be infinite; instead it would be just $1/(1 - R)$ times larger than the initial population. For example, one-child reproduction practices (R = 0.5) would only lead to a doubling of the total immortal population, because $1/(1 - 0.5) = 2$. In other words, a population of immortal reproducing organisms can grow indefinitely in time, [without necessarily growing] indefinitely in size, because . . . infinite geometric series converge when the absolute value of the common ratio (R) is less than one. The startling conclusion is that fears of overpopulation based on lay common sense and uneducated intuition are, in fact, grossly exaggerated." Vicki Glaser, "Interview with Leonid A. Gavrilov, Ph.D. and Natalia Gavrilova, Ph.D.," *Rejuvenation Research* 12, no. 5 (2009): 373; Leonid A. Gavrilov and Natalia S. Gavrilova, "Demographic Consequences of Defeating Aging," *Rejuvenation Research* 13, no. 2–3 (April 2010): 329–34.

10. Worldwatch Institute, "The State of Consumption Today," http://www.worldwatch.org/node/810.

11. Cohen, *How Many People Can the Earth Support?*; Angus and Butler, *Too Many People?*; Kenny, *Getting Better*.

12. One further factor may influence this equation. Individuals who expect to live 160 years might tend to adopt a quite different attitude toward long-term ecological policies than those who anticipate dying around age eighty. Knowing that you yourself will be there in person to experience the consequences of today's environmental practices may prove potently compelling as a motivator for espousing green values. While the concept of intergenerational responsibility offers a profound and laudable ideal, the tragic fact of the matter is that the society of the middle-term future remains, in the eyes of most people, little more than a dim abstraction. Nobody can really know what their great-great-great-great-grandchildren will be like. But the

image of you yourself—and your equally long-lived children—having to deal someday with dire water shortages or massive climate disasters, is far from abstract. It
may provide a badly needed impetus for adopting ambitious green strategies oriented toward the long haul. Longer-lived people may make greener citizens. See
Zey, *Ageless Nation*, chap. 8; Arrison, *100 Plus*, 63–64.

13. Elizabeth Abbott, *A History of Marriage* (New York: Seven Stories, 2010); Stephanie
Coontz, *Marriage, a History: From Obedience to Intimacy, or How Love Conquered Marriage* (New York: Penguin, 2005).

14. The writer Drew Magary, in his entertaining sci-fi novel *The Postmortal*, envisions a
world in which people can get "The Cure" and halt their biological aging altogether.
One consequence, in Magary's estimation, is an end to the principle of "til death do
us part." Instead, many pairs of lovers enter voluntarily into fixed nuptial contracts
with thirty- or forty-year terms. In my opinion, however, such an outcome seems
unlikely. Most people hold a more romantic conception of love, which precludes
looking into their partner's eyes and saying, "You are the love of my life—for the
next thirty-five years." The statistics on prenuptial agreements in today's society
bear this out: only 3 percent of marriages are bound by such documents. Everyone
else takes a more hopeful stance of "all in." Drew Magary, *The Postmortal* (New York:
Penguin, 2011); Laura Petrecca, "Prenuptial agreements: Unromantic, but important," *USA Today*, March 11, 2010; Coontz, *Marriage: A History*.

15. Zey, *Ageless Nation*, chap. 5.

16. Arrison, *100 Plus*, chap. 5.

17. Zey, *Ageless Nation*, chap. 5.

18. It is quite possible that a longer-lived population will be, overall, a more prosperous
population. Economists who study productivity and quality of life have concluded
that even relatively small gains in the state of health and life expectancy of a nation's
citizenry correlate with significantly increased overall prosperity. Partly this is because healthy people are obviously more likely to be productive than unhealthy
ones, but another key factor has to do with the accumulation of education and experience. Longer-lived people are more likely, on average, to acquire higher levels of
education than short-lived ones; they are also able to bring to their jobs the myriad
lessons learned through years of hands-on practice. See the extensive bibliography at
the end of Kevin Murphy and Robert Topel, "The Value of Health and Longevity,"
Journal of Political Economy 114, no. 5 (October 2006): 871–904.

19. Zey, *Ageless Nation*, chaps. 4, 8.

20. Ibid., 201.

21. This scenario assumes, of course, that universities will still exist a hundred years
from now—which is far from clear. But regardless of the physical setting, the basic
principle of reprising a life phase that is primarily dedicated to education would
still apply.

22. Zey, *Ageless Nation*, chap. 8; Arrison, *100 Plus*, chap. 6.

23. See the excellent discussion in Jay Rosenberg, "Reassessing Immortality: The Mak-
ropoulos Case Revisited," in *The Good, the Right, Life and Death: Essays in Honor of
Fred Feldman*, ed. Kris McDaniel, Jason Raibley, Richard Feldman, and Michael
Zimmerman (Burlington, VT: Ashgate, 2006), 227–40.

Fifteen: New Sounds for the Old Guitar

1. Steve Martin, Carl Reiner, and George Gipe, *The Man with Two Brains* (Warner Brothers, 1983).

2. Over the past two decades, scholars from a variety of disciplines have explored the nature of gender—its relation to anatomical features and biological processes, the meanings it has held across different cultures and historical epochs, the roles it has played in supporting a range of social hierarchies and power structures. They have made a persuasive case that the relation between our biological bodies and our sense of gender is considerably more pliable than most people tend to think. In 1992, for example, the historian Thomas Laqueur argued in his book *Making Sex* that the concept of "male" and "female" as two distinct sexes is actually a fairly recent histor-ical development: before the eighteenth-century Enlightenment, females were cate-gorized in Western culture as an (inferior) subset of a single human sex, of which males were the highest incarnation. The literature on the subject is vast. A good place to start is with the following works and their bibliographies: Thomas Laqueur, *Making Sex: Body and Gender from the Greeks to Freud* (Cambridge, MA: Harvard University Press, 1992); Anne Fausto-Sterling, *Sexing the Body: Gender Politics and the Construction of Sexuality* (New York: Basic, 2000); Judith Butler, *Gender Trouble: Femi-nism and the Subversion of Identity* (New York: Routledge, 2006); Margaret Lock and Judith Farquhar, eds., *Beyond the Body Proper: Reading the Anthropology of Material Life* (Durham, NC: Duke University Press, 2007); Ian Hacking, *The Social Construction of What?* (Cambridge, MA: Harvard University Press, 2000).

3. I encountered this intriguing concept on a website titled *Future of Sex*, http://futureofsex.net/.

4. Regina Lynn, "Ins and Outs of Teledildonics," *Wired*, September 24, 2004; David Levy, *Love and Sex with Robots: The Evolution of Human-Robot Relationships* (New York: Harper, 2007), 266–68.

5. Levy, *Love and Sex with Robots*.

6. Here the concept of the "uncanny valley" may prove relevant. Researchers have dis-covered that most people have no problem relating to robots if they are very unlike humans in appearance, but that a reaction of revulsion tends to result if the robot looks *almost but not quite* like a human. If the robot looks too much like a human, but has discernible artificial features such as excessive sheen on its skin or excessively symmetrical facial features, then most people find the similarity deeply unsettling. See, for example, Masahiro Mori, "The Uncanny Valley," *IEEE Robotics & Automa-tion Magazine* 19, no. 2 (1970/2012): 98–100, doi:10.1109/MRA.2012.2192811.

7. Scott Gelfand and John Shook, eds., *Ectogenesis: Artificial Womb Technology and the Future of Human Reproduction* (Amsterdam: Rodopi, 2006); Frida Simonstein, ed., *Reprogen-Ethics and the Future of Gender* (New York: Springer, 2009).

8. Canwest News Service, "Miracle Child," Canada.com, February 11, 2006, http://www.canada.com/topics/bodyandhealth/story.html?id=db8f33ab-33e9-429f-bedc-b6ca80f61bdc.

9. See the essays in Gelfand and Shook, eds., *Ectogenesis*; and Simonstein, ed., *Reprogen-Ethics and the Future of Gender*.

10. Ibid.

11. Simonstein, ed., *Reprogen-Ethics and the Future of Gender*.

12. Josh Schonwald, *The Taste of Tomorrow: Dispatches from the Future of Food* (New York: Harper Collins, 2012), chap. 19.

13. Ibid.

14. Noah Shachtman, "DARPA Offers No Food for Thought," *Wired*, February 17, 2004; Michael Belfiore, *The Department of Mad Scientists: How DARPA Is Remaking Our World, from the Internet to Artificial Limbs* (New York: Harper, 2009).

15. Schonwald, *The Taste of Tomorrow*, chap. 18.

16. Robert Freitas, quoted in "2010 Solutions for a Better Future," ed. Patrick Tucker, *The Futurist*, January-February 2010, http://www.wfs.org/Dec09-Jan10/solutions .htm. See Freitas's webpage, http://www.rfreitas.com/. See also Schonwald, *The Taste of Tomorrow*, 264–70.

17. Freitas, quoted in "2010 Solutions for a Better Future."

18. David Holtzman, *Privacy Lost: How Technology Is Endangering Your Privacy* (San Francisco: Jossey-Bass, 2006), xii–xiv.

19. The information we generate falls into two broad categories, which can be thought of as "deliberate data" and "inadvertent data." The former includes all the digital items that we intentionally set about creating and recording for the myriad purposes that occupy us every day, from work to leisure: e-mails and text messages, spreadsheets and word documents, photos and videos, social media entries. The latter comprises all the data we generate without realizing it (or which, even if we do realize we are generating it, comes into being as a side effect of our action rather than as its primary purpose). Thus, for example, inadvertent data arise when you go to a website online and your computer leaves a cookie on the site's server to mark your visit; when you call someone on a cell phone and the exact location of where you are standing is logged via GPS and via the ID of the cell phone tower through which your telecommunication was routed; when you search for something on Google and your search is stored (indefinitely) on Google's mainframes; when you go to the doctor for an eye infection and your visit becomes a permanent digital record owned by your insurance company; when a murderer unwittingly sheds a piece of his own skin or hair near the body of his victim, thereby providing investigators with a DNA profile that reveals his identity. Holtzman, *Privacy Lost*; David Brin, *The Transparent Society: Will Technology Force Us to Choose Between Privacy and Freedom?* (New York: Basic, 1998).

20. Heather Kelly, "Self-Driving Cars Now Legal in California," CNN, September 27, 2012, http://www.cnn.com/2012/09/25/tech/innovation/self-driving-car-california /index.html. To clarify the headline: a human co-driver is still legally required to be on board.

21. See "Wearable Technology," *Wikipedia*, http://en.wikipedia.org/wiki/ Wearable_technology#Prototypes.

22. Cynthia Breazeal, *Designing Sociable Robots* (Cambridge, MA: MIT Press, 2002).

23. Richard Harper, ed., *Inside the Smart Home* (New York: Springer, 2003); Danny Briere and Pat Hurley, *Smart Homes for Dummies* (Hoboken, NJ: Wiley, 2007).

24. Serkan Toto, "Japanese Toilet Analyzes Stool, Beams Results to Cell Phones via Personalized URLs," *TechCrunch*, April 1, 2009, http://techcrunch.com/2009/04/01 /japanese-toilet-analyzes-stool-beams-results-to-cell-phones-via-personalized-urls/.

25. Adam Greenfield, *Everyware: The Dawning Age of Ubiquitous Computing* (Berkeley, CA: New Riders, 2006), 18–19.

26. Robert Scoble and Shel Israel, *Age of Context: Mobile, Sensors, Data, and the Future of Privacy* (Patrick Brewster Press, 2013); David Rose, *Enchanted Objects: Design, Human Desire, and the Internet of Things* (New York: Scribner, 2014).

27. Holtzman, *Privacy Lost*, chaps. 12–13; Lawrence Lessig, *Code: Version 2.0* (New York: Basic, 2006), chap. 11.

28. Richard Wagner, *The Art-Work of the Future*, trans. William Ellis (London: Kegan Paul, 1895).

29. In 2008, Japan's ATR Computational Laboratories used fMRI to retrieve from the brain of a human subject a visual image he was perceiving—the letters spelling the word "neuron" (see chapter 3). Some observers immediately made the giddy leap into an impending world of mind reading and perhaps the monitoring of our brain's activity during sleep. "By applying this technology," noted one of the project scientists, "it may become possible to record and replay subjective images that people perceive in their dreams." Quoted in "Have You Been Dreaming of a White Christmas? Scientists Could Soon Watch It on a Screen," *Daily Mail Online*, December 11, 2008, http://www.dailymail.co.uk/sciencetech/article-1093770/Have-dreaming -white-Christmas-Scientists-soon-watch-screen.html.

30. In his most influential work, *Understanding Media* (1964), Marshall McLuhan argued that the communication media prevalent in a given epoch powerfully shape and constrain the way people think, interact, and organize their lives. Thus, for example, a predominantly oral culture will exhibit very different concepts of selfhood, social boundaries, temporality, and artistic possibilities than a print culture—which in turn will differ in equally significant ways from an electronic culture (radio or TV). McLuhan's point is frequently summed up with the aphorism, "The medium is the message"—by which he meant that the *form* itself through which information is communicated exerts a transformative effect on those who adopt it, above and beyond the particular *content* the information conveys. Marshall McLuhan, *Understanding Media: The Extensions of Man* (repr.; Cambridge, MA: MIT Press, 1994).

31. M. T. Anderson, *Feed* (Cambridge, MA: Candlewick Press, 2002).

32. Nicholas Carr, *The Shallows: What the Internet Is Doing to Our Brains* (New York: Norton, 2011).

33. Lessig, *Code: Version 2.0*.

34. Sherry Turkle, *Alone Together: Why We Expect More from Technology and Less from Each Other* (New York: Basic, 2011).

35. Jaron Lanier, *You Are Not a Gadget: A Manifesto* (New York: Knopf, 2010), 4.

36. See, on this book's companion website, http://www.ourgrandchildrenredesigned.org, the books by Tucker and Danzig; Gray; Moreno; and Singer in the online bibliography section on law, government, and military applications.

37. See the books by Tucker and Danzig; Koblentz; Gray; Guillemin; Alibek and Handelman; Miller, Broad, and Engelberg; and Luther, Lebeda, and Korch in the online bibliography section on law, government, and military applications.

38. Michael Belfiore, *The Department of Mad Scientists: How DARPA Is Remaking Our World, from the Internet to Artificial Limbs* (New York: Harper, 2009); Joel Garreau, *Radical Evolution: The Promise and Peril of Enhancing Our Minds, Our Bodies—and What It Means to Be Human* (New York: Doubleday, 2004), 22–44; Jonathan Moreno, *Mind Wars: Brain Research and National Defense* (New York: Dana, 2006).

39. The text quoted is from a version of the DARPA web page accessed in 2008. The last sentence of the quoted paragraph has recently been toned down a bit: "Since the very beginning, DARPA has been the place for people with innovative ideas that lead to groundbreaking discoveries." DARPA's website provides plenty of information about its many programs. See http://www.darpa.mil/About.aspx.

40. Belfiore, *The Department of Mad Scientists*; Garreau, *Radical Evolution*, 22–44; Moreno, *Mind Wars*.

41. I derive the five categories that follow from the richly informative DARPA website and from the following other sources: Belfiore, *The Department of Mad Scientists*; Garreau, *Radical Evolution*, 22–44; Moreno, *Mind Wars*; Singer, *Wired for War*. See also the extensive DARPA overview provided to the US Congress in February 2007: "DARPA: Bridging the Gap, Powered by Ideas," http://oai.dtic.mil/oai/oai?verb =getRecord&metadataPrefix=html&identifier=ADA510795. See also the military-themed sections in Mihail Roco and William Sims Bainbridge, eds., *Converging Technologies for Improving Human Performance: Nanotechnology, Biotechnology, Information Technology, and Cognitive Science* (Dordrecht, Holland: Kluwer, 2003); William Bainbridge and Mihail C. Roco, eds., *Managing Nano-Bio-Info-Cogno Innovations: Converging Technologies in Society* (New York: Springer, 2006); and Bruce Katz, *Neuro-engineering the Future: Virtual Minds and the Creation of Immortality* (Sudbury, MA: Infinity Science Press, 2008).

42. In 2009, Berkeley Bionics introduced a wearable exoskeleton, powered by a portable backpack, that allows people to carry two-hundred-pound loads for long treks over rugged terrain. The lightweight and surprisingly minimalist-looking device consists of two robotic legs controlled by a portable microcomputer. Worn over the human's legs like a spindly external crutch, the exoskeleton senses and accompanies a person's every move, allowing remarkable flexibility and freedom of motion. It also absorbs most of the load-bearing stress associated with walking or running, allowing a person carrying a bone-crushingly heavy backpack to experience less fatigue than someone walking unburdened. At journey's end, the device snaps off and dismantles with a few quick flicks of its mechanical catches. Berkeley Bionics, founded in 2005, changed its name in 2011 to Ekso Bionics, http://www.eksobionics.com/about-us. For a video of the exoskeleton in action, see "Berkeley Bionics Human Exoskeleton," YouTube, http://www.youtube.com /watch?v=EdK2y3lphmE.

Sixteen: Why Extreme Modifications Should Be Postponed

1. Giovanni Pico della Mirandola, *On the Dignity of Man*, trans. Charles Wallis (Cambridge, MA: Hackett, 1998), 5.

2. A pioneering work in this regard was Jonathan Glover, *What Sort of People Should There Be?* (New York: Penguin, 1984). For more recent works, see especially the books by Baillie; Burley and Harris; Chapman and Frankel; DeGrazia; Glannon; Heyd; McGee; Newman; and Stock in the online bibliography section on genetics on this book's companion website, http://www.ourgrandchildrenredesigned.org.

3. Freeman Dyson ponders such issues with his customary perspicacity in *Imagined Worlds* (Cambridge, MA: Harvard University Press, 1997), chap. 4. See also the final chapters of Derek Parfit, *On What Matters*, vol. 2 (New York: Oxford University

Press, 2011); and Arthur C. Clarke, *Profiles of the Future: An Inquiry into the Limits of the Possible* (London: Indigo, 1999), chaps. 17–19.

4. See also the works under the online bibliography heading of "Justice, ethics" on this book's companion website.

5. A concise overview of the challenges posed by international regulation is presented in Jonathan Tucker and Richard Danzig, eds., *Innovation, Dual Use, and Security: Managing the Risks of Emerging Biological and Chemical Technologies* (Cambridge, MA: MIT Press, 2012), 305–40; see also Graeme Hodge, Diana Bowman, and Andrew Maynard, eds., *International Handbook on Regulating Nanotechnologies* (Cheltenham, UK: Edward Elgar, 2010), chaps. 23 and 24.

6. This conclusion, it is worth noting, does not amount to a form of "technological determinism." I am *not* claiming that enhancement technologies possess an implacable momentum of their own, and that we humans have no choice but to accept them into our social world, because they are coming whether we like it or not. Rather, I am arguing that our society could indeed block the development of all such technologies if it absolutely wanted to—but that the social, economic, scientific, and political costs of doing so would necessarily be very high. It would require a series of painful trade-offs that most citizens would probably, in the end, choose to reject.

7. Two excellent discussions of radical enhancement are Nicholas Agar, *Humanity's End: Why We Should Reject Radical Enhancement* (Cambridge, MA: MIT Press, 2010); and Allen Buchanan, *Beyond Humanity? The Ethics of Biomedical Enhancement* (New York: Oxford University Press, 2011). Both Agar and Buchanan tend to use the term "radical enhancement" to describe modifications that I categorize for the most part under the heading of "mid-level modifications." My category of "high-level enhancement" is reserved for the far more profoundly transformed entities described by Ray Kurzweil, Hans Moravec, and other transhumanists.

8. The full description runs as follows: "Many transhumanists wish to follow life paths which would, sooner or later, require growing into posthuman persons: they yearn to reach intellectual heights as far above any current human genius as humans are above other primates . . . ; to have unlimited youth and vigor; to exercise control over their own desires, moods, and mental states; to be able to avoid feeling tired, hateful, or irritated about petty things; to have an increased capacity for pleasure, love, artistic appreciation, and serenity; to experience novel states of consciousness that current human brains cannot access. . . . Some posthumans may find it advantageous to jettison their bodies altogether and live as information patterns on vast super-fast computer networks. Their minds may be not only more powerful than ours but may also employ different cognitive architectures or include new sensory modalities that enable greater participation in their virtual reality settings." Humanity+ web site, http://humanityplus.org/philosophy/transhumanist-faq/#answer_20.

9. One of the best books available on human personhood is Christian Smith, *What Is a Person? Rethinking Humanity, Social Life, and the Moral Good from the Person Up* (Chicago: University of Chicago Press, 2010).

10. An illuminating discussion of this necessary focalization of consciousness is given in Thomas Metzinger, *The Ego Tunnel: The Science of the Mind and the Myth of the Self* (New York: Basic, 2009). See also Douglas Hofstadter and Daniel Dennett, *The Mind's I: Fantasies and Reflections on Self and Soul* (New York: Basic, 1981).

11. Derek Parfit, *Reasons and Persons* (New York: Oxford University Press, 1986), chap. 12.

12. Smith, *What Is a Person?*; Metzinger, *The Ego Tunnel*; Parfit, *Reasons and Persons*; Hofstadter and Dennett, *The Mind's I*; Marvin Minsky, *The Emotion Machine: Commonsense Thinking, Artificial Intelligence, and the Future of the Human Mind* (New York: Simon & Schuster, 2006).

13. In his book, *The Singularity Is Near*, for example, Ray Kurzweil seeks to reassure his readers that all the radical changes coming down the pike will still leave us recognizably human in the fundamentals of our being. "The intelligence that will emerge [after the Singularity]," he says, "will continue to represent the human civilization, which is already a human-machine civilization. In other words, future machines will be human, even if they are not biological. This will be the next step in evolution. . . . Most of the intelligence of our civilization will ultimately be nonbiological. By the end of this century, it will be trillions of trillions of times more powerful than human intelligence. However, to address often-expressed concerns, this does not imply the end of biological intelligence, even if it is thrown from its perch of evolutionary superiority. Even the nonbiological forms will be derived from biological design. Our civilization will remain human—indeed, in many ways it will be more exemplary of what we regard as human than it is today, although our understanding of the term will move beyond its biological origins." Ray Kurzweil, *The Singularity Is Near* (New York: Viking, 2005), 30.

14. See appendix K on this book's companion website for a discussion of the transhumanists' idea of downloading human consciousness into a machine body.

15. Tom Beauchamp, *Intending Death: The Ethics of Assisted Suicide and Euthanasia* (Zug, Switzerland: Pearson, 1995); Demetra Pappas, *The Euthanasia/Assisted-Suicide Debate* (Santa Barbara, CA: Greenwood, 2012).

16. Some opponents of assisted suicide argue that such a deed harms all humans by implicitly lessening the value of human life. This is an argument that merits serious consideration, but it is based on an assessment of an *indirect* form of harm that is allegedly being caused to society, not a direct and clear danger.

17. See the discussion in Julian Morris, *Rethinking Risk and the Precautionary Principle* (Oxford, UK: Butterworth/Heinemann, 2000); Cass Sunstein, *Laws of Fear: Beyond the Precautionary Principle* (New York: Cambridge University Press, 2005); Thomas Coleman, *A Practical Guide to Risk Management* (Charlottesville, VA: CFA Institute, 2011); Kerry Whiteside, *Precautionary Politics: Principle and Practice in Confronting Environmental Risk* (Cambridge, MA: MIT Press, 2006).

18. In appendix P online, on this book's companion website, I discuss how this line of reasoning also applies to the creation of human-level forms of artificial intelligence. See Nick Bostrom, *Superintelligence: Paths, Dangers, Strategies* (New York: Oxford University Press, 2014); Stuart Armstrong, *Smarter Than Us: The Rise of Machine Intelligence* (Berkeley, CA: MIRI, 2014); Hugo De Garis, *The Artilect War: Cosmists vs. Terrans: A Bitter Controversy Concerning Whether Humanity Should Build Godlike Massively Intelligent Machines* (Ottawa, Canada: ETC, 2005); and James Barrat, *Our Final Invention: Artificial Intelligence and the End of the Human Era* (New York: Thomas Dunne, 2013).

19. Extensions of the human health span may not need to be subjected to legal regulation at all, since the gradual and incremental nature of their development will yield the same practical result as a phased moratorium.

20. See appendix A online, on this book's companion website, where I describe the diffi-
    culties today's scholars are already encountering in defining what constitutes an
    enhancement.
21. Some citizens would no doubt argue that the moral and economic cost of such a
    moratorium is worse than the evils it is designed to prevent. Defenders of the mora-
    torium would then need to show that—at least, for certain extreme categories of
    bioenhancements—the danger is sufficiently frightful to outweigh those costs.
22. The literature on this subject is vast. See, for example, Michael Bess, *Realism, Utopia,
    and the Mushroom Cloud: Four Activist Intellectuals and their Strategies for Peace* (Chicago:
    University of Chicago Press, 1993); David Kearn, *Great Power Security Cooperation:
    Arms Control and the Challenge of Technological Change* (Lanham, MD: Lexington, 2014);
    Jonathan Tucker and Richard Danzig, eds., *Innovation, Dual Use, and Security: Manag-
    ing the Risks of Emerging Biological and Chemical Technologies* (Cambridge, MA: MIT
    Press, 2012); Graeme Hodge, Diana Bowman, and Andrew Maynard, eds., *Interna-
    tional Handbook on Regulating Nanotechnologies* (Cheltenham, UK: Edward Elgar, 2010).
23. Along similar lines, the AI researcher Hugo de Garis has suggested that these kinds
    of preemptive wars might be triggered by the advent of human-level artificial intel-
    lects. If one nation or group of nations were rapidly developing a capability to build
    human-level AIs, then other nations might regard this as a sufficiently grave techno-
    logical threat to justify a preemptive military action. See Garis, *The Artilect War*.

Seventeen: Humane Values in a World of Moderate Enhancements

 1. Frances McDormand in Joel Coen and Ethan Coen, *Burn After Reading* (Relativity
    Media, Studio Canal, Working Title Films, 2008).
 2. For a detailed balance sheet of the likely advantages and drawbacks of bioenhance-
    ments, see appendix Q on this book's companion website, http://www.ourgrand
    childrenredesigned.org.
 3. Francis Castles, *The Social Democratic Image of Society: A Study of the Achievements and
    Origins of Scandinavian Social Democracy in Comparative Perspective* (New York: Rout-
    ledge, 2009); Francis Sejersted, *The Age of Social Democracy: Norway and Sweden in the
    Twentieth Century*, trans. Richard Daly (Princeton, NJ: Princeton University Press,
    2011); M. Donald Hancock, ed., *Politics in Europe*, 5th ed. (Washington, DC: CQ
    Press, 2011); Mary Hilson, *The Nordic Model: Scandinavia since 1945* (London: Reak-
    tion, 2008).
 4. See the discussion regarding the eugenics movement in chapter 4.
 5. Some might argue that there can be no such thing as "neutrality" in choosing which
    enhancements to subsidize and which ones to disallow. Under this view, *any* restric-
    tive criteria adopted by the government in laying out enhancement options for its
    citizens—even the most putatively broad-minded and generous set of options—
    would unavoidably constitute, ipso facto, a tacit form of eugenics. This position has
    merit, of course, but a pragmatist would reply that the government ultimately has no
    choice. Its role as subsidizer and regulator leaves no other option but to settle on
    some set of restrictive criteria. No government could afford the cost of indiscrimi-
    nately offering its citizens any and all enhancements they dream up. Some enhance-
    ments, moreover, would need to be excluded because they would demonstrably
    undermine the safety or welfare of other citizens. These considerations underscore

the delicate nature of the task that would face government officials in laying out the criteria for subsidizing and regulating enhancements. They would need to engage in a difficult series of ongoing trade-offs among multiple values: maximizing the range of options open to citizens; making sure the technologies are safe; controlling costs; and ensuring that citizens' enhancement choices do not endanger the welfare of other citizens. See Nicholas Agar, *Liberal Eugenics: In Defence of Human Enhancement* (Malden, MA: Wiley-Blackwell, 2004).

6. David Hulme, *Global Poverty: How Global Governance Is Failing the Poor* (New York: Routledge, 2010).

7. Richard Benedick, *Ozone Diplomacy: New Directions in Safeguarding the Planet* (Cambridge, MA: Harvard University Press, 1998).

8. The Marshall Plan arguably constitutes another powerful precedent. Over a four-year period, from 1948 to 1952, the US government gave out some $12 billion in aid to European nations that were struggling to get their economies restarted after the devastation of World War II. This was a significant amount—about one-third of a single year's federal budget. The man behind the initiative, Secretary of State George C. Marshall, reasoned as follows: the European peoples in 1945 and 1946 were literally starving to death; their economies were floundering, their political institutions teetering unstably. Under these conditions, Marshall knew, communism and other extremist ideologies would find fertile ground. He concluded that the best way to forestall renewed turmoil in Europe was for the United States to bankroll a swift European economic recovery. The gamble paid off handsomely. Marshall's intervention took one of the world's potentially most dynamic and prosperous regions, and firmly channeled it down a course that kept it aligned with American values and institutions: in the long run, all the European nations that participated in the Marshall Plan would remain democratic and capitalist. Greg Behrman, *The Most Noble Adventure: The Marshall Plan and How America Helped Rebuild Europe* (New York: Free Press, 2008).

9. Such an initiative is described in detail in Jeffrey Sachs, *The End of Poverty: Economic Possibilities for Our Time* (New York: Penguin, 2006). See also Abhijit Banerjee and Esther Duflo, *Poor Economics: A Radical Rethinking of the Way to Fight Global Poverty* (New York: PublicAffairs, 2011).

10. Behrman, *The Most Noble Adventure*.

11. Ed Cray, *General of the Army: George C. Marshall, Soldier and Statesman* (New York: Cooper Square, 2000), chap. 35.

12. Joseph Montville, *Conflict and Peacemaking in Multiethnic Societies* (Lanham, MD: Lexington, 1991); Frances Stewart, ed., *Horizontal Inequalities and Conflict: Understanding Group Violence in Multiethnic Societies* (New York: Palgrave-Macmillan, 2010); Karen Barkey and Mark Von Hagen, eds., *After Empire: Multiethnic Societies and Nation-building: The Soviet Union and the Russian, Ottoman, and Habsburg Empires* (Boulder, CO: Westview Press, 1997).

13. William Schabas, *The UN International Criminal Tribunals: The Former Yugoslavia, Rwanda and Sierra Leone* (New York: Cambridge University Press, 2006).

14. Three excellent books on this topic are Parker Palmer, *Healing the Heart of Democracy: The Courage to Create a Politics Worthy of the Human Spirit* (San Francisco: Jossey-Bass, 2011); Jonathan Haidt, *The Righteous Mind: Why Good People Are Divided*

by Politics and Religion (New York: Pantheon, 2012); and Robert Talisse, *Democracy and Moral Conflict* (New York: Cambridge University Press, 2009).

15. Maha Shuayb, *Rethinking Education for Social Cohesion: International Case Studies* (New York: Palgrave-Macmillan, 2012); Bradley Levinson and Doyle Stevick, *Reimagining Civic Education: How Diverse Societies Form Democratic Citizens* (Lanham, MD: Rowman & Littlefield, 2007); Norman Nie, Jane Junn, and Kenneth Stehlik-Barry, *Education and Democratic Citizenship in America* (Chicago: University of Chicago Press, 1996).

16. The work of the Australian artist Patricia Piccinini offers an excellent exemplar of the kind of thinking that will be required. Over the past couple decades, her sculptures, drawings, and paintings have systematically explored the boundaries among humans, animals, and machines, reconfiguring them in startling ways. The hybrid creatures she has envisioned—part human, part cow, part manatee, for example—are deliberately shown in circumstances that underscore their underlying personhood and dignity. Her fused animal-persons are often deeply unsettling in their alien appearance, but always, at the same time, clearly recognizable as possessing human attributes such as gentleness, timidity, vulnerability, motherliness, curiosity. In Piccinini's world, the "other" is definitely *very* other, but she relentlessly shows us that underneath the surface strangeness, this creature we are encountering shares many of the qualities we value most highly in ourselves. One leaves an exhibit of her work feeling that "aliens" need not necessarily be as disturbing as one initially felt—and that we should open our hearts and minds more readily to beings who are in many ways radically unlike us. The schoolchildren of the year 2100 would benefit greatly from being shown such boundary-challenging art and discussing what it means to them. "Seek the underlying commonalities"—this could become the watchword for this new era of civic education. See Piccinini's website, http://www .patriciapiccinini.net/. See also Donna Haraway, "Speculative Fabulations for Technoculture Generations: Taking Care of Unexpected Country," on Piccinini's website; Donna Haraway, *When Species Meet* (Minneapolis: University of Minnesota Press, 2008), chap. 12; and Kim Toffoletti, *Cyborgs and Barbie Dolls: Feminism, Popular Culture, and the Posthuman Body* (London: I. B. Tauris, 2007).

17. In such a world, defenders of animal rights would continue to safeguard the interests of all animal species, but they would also need to articulate new guidelines aimed at protecting the integrity of existing species boundaries. I argued in chapter 9 that entirely new forms of suffering might come into being as a result of human tampering with animal genomes. Therefore, the jurisprudence of animal rights will need to erect strong barriers to prevent irresponsible or cruel modifications of animals from being carried out. The creation of transgenic pets, for example, may strike some people as a relatively benign genetic intervention; yet it is not a self-evidently legitimate act. Moral and legal thinkers of the coming century will need to reflect long and hard about the implications of such wanton mixings. See chapters 11–12 and 20–24 in Tom Beauchamp and R. G. Frey, eds., *The Oxford Handbook of Animal Ethics* (New York: Oxford University Press, 2011); Donna Haraway, *Primate Visions: Gender, Race, and Nature in the World of Modern Science* (New York: Routledge, 1989); Haraway, *When Species Meet*; Cass Sunstein and Martha C. Nussbaum, eds., *Animal Rights: Current Debates and New Directions* (New York: Oxford University Press,

2004); Kelly Oliver, *Animal Lessons: How They Teach Us to Be Human* (New York: Columbia University Press, 2009).

18. Raymond Gibbs Jr., *Embodiment and Cognitive Science* (New York: Cambridge University Press, 2005); Antonio Damasio, *Self Comes to Mind: Constructing the Conscious Brain* (New York: Pantheon, 2010); Robert Kane, ed., *The Oxford Handbook of Free Will*, 2nd ed. (New York: Oxford University Press, 2011); John Perry, ed., *Personal Identity* (Oakland: University of California Press, 1975); Debra Matthews, Hilary Bok, and Peter Rabins, eds., *Personal Identity and Fractured Selves: Perspectives from Philosophy, Ethics, and Neuroscience* (Baltimore: Johns Hopkins University Press, 2009).

19. In our present society, alas, the trend appears to be going the opposite way, with many citizens perfectly willing to assent to greater governmental surveillance in the name of fighting crime or preempting terrorist attacks: "I'm not doing anything wrong, so it's fine for my government to spy on me." See, for example, the discussion in Glenn Greenwald, *No Place to Hide: Edward Snowden, the NSA, and the US Surveillance State* (New York: Metropolitan, 2014).

20. See the detailed critique of such economists' views in Michael Sandel, *What Money Can't Buy: The Moral Limits of Markets* (New York: Farrar, Straus and Giroux, 2012); and Debra Satz, *Why Some Things Should Not Be for Sale: The Moral Limits of Markets* (New York: Oxford University Press, 2010).

21. Sandel, *What Money Can't Buy*; Satz, *Why Some Things Should Not Be for Sale*. I draw heavily on both of these excellent works in the account that follows.

22. See Sandel, *What Money Can't Buy*, chap. 3.

23. Ibid., chap. 2.

24. Satz, *Why Some Things Should Not Be for Sale*, chap. 4; Sandel, *What Money Can't Buy*, chap. 5.

25. Michael Zimmerman, *The Nature of Intrinsic Value* (Lanham, MD: Rowman & Littlefield, 2001); Noah Lemos, *Intrinsic Value: Concept and Warrant* (New York: Cambridge University Press, 2009).

26. Arlie Hochschild, *The Managed Heart: Commercialization of Human Feeling*, 2nd ed. (Oakland: University of California Press, 2012); Martha Ertman and Joan Williams, *Rethinking Commodification: Cases and Readings in Law and Culture* (New York: New York University Press, 2005); Zygmunt Bauman, *Consuming Life* (Cambridge, MA: Polity, 2007); see also the works by Sandel; Satz; Zimmerman; and Lemos, cited above.

27. A sci-fi novel that examines this premise is M. T. Anderson, *Feed* (Cambridge, MA: Candlewick Press, 2002).

28. The ethics of such campaigns to mold public behavior is taken up in Sarah Conly, *Against Autonomy: Justifying Coercive Paternalism* (New York: Cambridge University Press, 2012); and in a review of Conly's book by Cass Sunstein, "It's for Your Own Good!" *New York Review of Books*, March 7, 2013, 8–11.

29. Alan Schwarz, "Drowned in a Stream of Prescriptions," *New York Times*, February 2, 2013.

30. This is not to deny that ADHD is a real disorder with potentially devastating consequences for those who suffer from it, or that drugs like Ritalin and Adderall can make a huge difference for many individuals who are severely afflicted by the

condition. Still, our society has become too nonchalant about such diagnoses and too liberal in general in dispensing pills as a solution to all manner of such ethereal and subtle complaints. See Peter Conrad, *The Medicalization of Society: On the Transformation of Human Conditions into Treatable Disorders* (Baltimore: Johns Hopkins University Press, 2007); Adele Clarke et al., eds., *Biomedicalization: Technoscience, Health, and Illness in the US* (Durham, NC: Duke University Press, 2010).

31. Ingmar Persson and Julian Savulescu, *Unfit for the Future: The Need for Moral Enhancement* (New York: Oxford University Press, 2012); John Harris, "Moral Enhancement and Freedom," *Bioethics* 25, no. 2 (2011): 102–11; Lenny Moss, "Moral Molecules, Modern Selves, and Our 'Inner Tribe,'" *Hedgehog Review* 15, no. 1 (Spring 2013): 19–33; Peter Singer and Agata Sagan, "Are We Ready for a 'Morality Pill'?" *New York Times*, January 28, 2012, http://opinionator.blogs.nytimes.com/2012/01/28/are-we-ready-for-a-morality-pill/.

32. This is one of the key factors in the collapse of civilizations that Jared Diamond identifies in *Collapse: How Societies Choose to Fail or Succeed* (New York: Viking, 2005). Specifically, he points to three recurring problems: societies do things with disastrous consequences because they fail to anticipate a problem before it arrives; or because they fail to perceive or recognize a problem when it does arrive; or because, even after having recognized the existence of a problem, they fail to take cogent action to address it (see chap. 14).

33. Michael Pollan, *The Omnivore's Dilemma* (New York: Penguin, 2006); Steve Striffler, *Chicken: The Dangerous Transformation of America's Favorite Food* (New Haven, CT: Yale University Press, 2005).

34. If pressed by a reader who has never read anything by Berry to recommend a single short piece that gives a good sense of his worldview, I would point to the remarkable sixty-page short story, "Pray Without Ceasing," first published in 1992, and available in Wendell Berry, *Fidelity: Five Stories* (New York: Pantheon, 1992). Three fine books on Berry are Matthew Bonzo and Michael Stevens, *Wendell Berry and the Cultivation of Life* (Grand Rapids, MI: Brazos, 2008); Mark Mitchell and Nathan Schueter, eds., *The Humane Vision of Wendell Berry* (Wilmington, DE: ISI, 2011); Fritz Oehlschaeger, *The Achievement of Wendell Berry: The Hard History of Love* (Lexington: University Press of Kentucky, 2011).

35. This insight about groundedness in the present moment is, of course, not Berry's alone, but runs like a common thread through much of the literature on spirituality and the nature of a good life. A sampling of works that deal powerfully with this theme: Leo Tolstoy, *The Death of Ivan Ilych*, trans. Aylmer Maude (New York: Signet, 1960); Thich Nhat Hanh, *Being Peace* (Berkeley, CA: Parallax, 1987); Shunryu Suzuki, *Zen Mind, Beginner's Mind* (New York: Weatherhill, 1970); Jack Kornfield, *A Path With Heart* (New York: Bantam, 1993); Steven Levine, *A Gradual Awakening* (New York: Anchor, 1979).

Eighteen: What You and I Can Do Today

1. Maurice Goldhaber, quoted in *Nuclear Physics in Retrospect*, ed. Roger Stuewer (Minneapolis: University of Minnesota Press, 1979), 107.

2. See the brilliant discussion of historical agency in William Sewell, *Logics of History: Social Theory and Social Transformation* (Chicago: University of Chicago Press, 2005),

especially chaps. 4 and 8; and Christian Smith, *What Is a Person? Rethinking Humanity, Social Life, and the Moral Good from the Person Up* (Chicago: University of Chicago Press, 2010), especially chaps. 4–6.

3. I draw heavily in the argument that follows on an earlier work of mine: Michael Bess, *The Light-Green Society: Ecology and Technological Modernity in France, 1960–2000* (Chicago: University of Chicago Press, 2003). For a wide range of up-to-date resources on the topic see the website of the American Society for Environmental History, http://aseh.net/teaching-research/environmental-history-bibliographies.

4. Bess, *The Light-Green Society*, chaps. 7–10.

5. Ibid., chap. 11.

6. Ibid., chaps. 11, 12, and conclusion.

7. In 2001, for example, President George W. Bush decided that research on human embryonic stem cells was morally unacceptable because it relied on the creation and killing of human embryos. Despite opposition from a variety of social and political constituencies that considered such research morally legitimate, Bush issued an executive order that severely constrained further stem cell research in the United States. As a result, research on embryonic stem cells by American scientists slowed down considerably from 2001 to 2009—when the newly elected president Barack Obama issued an order that countermanded the restrictions imposed by his predecessor. Most notably, President Bush did not need to go so far as to issue an outright ban on this kind of research: he merely withdrew all federal funding from it, and this proved sufficient to achieve his goal. To be sure, the president's powers were not unlimited, for scientists in other countries such as Sweden and the UK were able to proceed unimpeded with their own work on human embryonic stem cells. But Bush's impact was nonetheless considerable: he successfully reined in a particular form of research that he and his philosophical associates found objectionable. His action, moreover, ultimately incentivized US scientists to develop ingenious techniques for generating pluripotent stem cells without using any human embryos at all. The broader lesson implicit here, both with the response to Sputnik and the restriction on stem cells, is that vigorous governmental action can not only speed up or slow down specific kinds of scientific and technological innovation; it can also forcefully channel that innovation down paths it might not otherwise have taken. Leo Furcht and William Hoffman, *The Stem Cell Dilemma: The Scientific Breakthroughs, Ethical Concerns, Political Tensions, and Hope Surrounding Stem Cell Research*, 2nd ed. (New York: Arcade, 2011), chap. 3.

8. Ralph Nader, *Unsafe at Any Speed* (New York: Grossman, 1965); Patricia Marcello, *Ralph Nader: A Biography* (Santa Barbara, CA: Greenwood, 2004).

9. They pledged to restrict their experiments to organisms that would not be able to flourish and multiply outside the laboratory context, and adopted a variety of containment measures, taking care that the modified organisms could not spread beyond their laboratories. Certain types of highly pathogenic organisms were declared off-limits to experimentation. Donald Frederickson, "The First Twenty-Five Years After Asilomar," *Perspectives in Biology and Medicine* 44, no. 2 (Spring 2001): 170–82; Marcia Barinaga, "Asilomar Revisited: Lessons for Today," *Science* 287, no. 5458 (March 3, 2000): 1584–85; Donald Frederickson, "Asilomar and Recombinant DNA: The End of the Beginning," in *Biomedical Politics*, ed. Kathi Hanna (Washington, DC: National

Academy Press, 1991), 258–324; Paul Berg et al., "Summary Statement of the Asilomar Conference on Recombinant DNA Molecules," *Proceedings of the National Academy of Sciences* 72, no. 6 (June 1975): 1981–84.

10. Frederickson, "The First Twenty-Five Years After Asilomar," 170–82; Barinaga, "Asilomar Revisited: Lessons for Today."

11. Kerry Whiteside, *Precautionary Politics: Principle and Practice in Confronting Environmental Risk* (Cambridge, MA: MIT Press, 2006); Cass Sunstein, *Laws of Fear: Beyond the Precautionary Principle* (New York: Cambridge University Press, 2005); Thomas Coleman, *A Practical Guide to Risk Management* (Charlottesville, VA: CFA Institute, 2011); Julian Morris, *Rethinking Risk and the Precautionary Principle* (Oxford, UK: Butterworth/Heinemann, 2000).

12. Robert Talisse, *Democracy and Moral Conflict* (New York: Cambridge University Press, 2009).

13. Robert Hazen and James Trefil, *Science Matters: Achieving Scientific Literacy* (New York: Anchor, 2009); Philip Kitcher, *Science in a Democratic Society* (New York: Prometheus, 2011); Jonathan Moreno, *The Body Politic: The Battle over Science in America* (New York: Bellevue, 2011); American Association for the Advancement of Science, *Benchmarks for Science Literacy* (New York: Oxford University Press, 1994).

14. Moreno, *The Body Politic*.

15. Al Gore, *The Future: Six Drivers of Global Change* (New York: Random House, 2013); Rudi Volti, *Society and Technological Change*, 6th ed. (New York: Worth, 2009), chap. 18.

16. I borrow the term "lock-in" from Jaron Lanier, *You Are Not a Gadget: A Manifesto* (New York: Knopf, 2010), 7–16. The concept is similar (but not identical) to that of "technological momentum" described by Thomas Hughes, *American Genesis: A Century of Invention and Technological Enthusiasm* (New York: Viking, 1989). See also Merritt Roe Smith and Leo Marx, eds., *Does Technology Drive History? The Dilemma of Technological Determinism* (Cambridge, MA: MIT Press, 1994); Gary Marchant, Braden Allenby, and Joseph Herkert, eds., *The Growing Gap Between Emerging Technologies and Legal-Ethical Oversight* (New York: Springer, 2011).

17. Dennis Cheek, *Thinking Constructively About Science, Technology, and Society Education* (Albany: SUNY Press, 1992); Sheila Jasanoff et al., eds., *Handbook of Science and Technology Studies* (Thousand Oaks, CA: Sage, 1995); Deborah Johnson and Jameson Wetmore, eds., *Technology and Society: Building Our Sociotechnical Future* (Cambridge, MA: MIT Press, 2009).

18. Another constructive factor, along these lines, would be to incorporate "anticipatory ethics" into legislation governing science and technology. The Human Genome Project (1989–2003) and National Nanotechnology Initiative (2003–present) are among the most significant scientific and technological endeavors undertaken by the US government over the past thirty years. Both these projects allocated a portion of their funding specifically for "ELSI" research: systematic analysis of the ethical, legal, and social implications of these cutting-edge domains of science, both in the present day and for the foreseeable future. (Full disclosure: the research for this book was partly funded by a grant from the ELSI program of the National Human Genome Research Institute.) This kind of mandated linkage between the funding of science and the funding of research into its societal consequences makes excellent sense and

should be generalized to other applications of science and technology research whenever possible. See Victor K. McElheny, *Drawing the Map of Life: Inside the Human Genome Project* (New York: Basic, 2012); John Sargent, *The National Nanotechnology Initiative: Overview, Reauthorization, and Appropriations Issues* (CreateSpace, 2012); Jerrod Kleike, ed., *National Nanotechnology Initiative: Assessment and Recommendations* (New York: Nova Science, 2010). For the ELSI component of the Human Genome Project, see the project webpage, http://www.ornl.gov/sci/techresources/Human _Genome/elsi/elsi.shtml.

For the ELSI component of the Nanotechnology Initiative, see http://www .nano.gov/you/ethical-legal-issues.

19. See Moreno, *The Body Politic*; and Kitcher, *Science in a Democratic Society*.

20. Moreno, *The Body Politic*, especially chaps. 6 and 7.

21. Richard Sclove, "Reinventing Technology Assessment," *Issues in Science and Technology* 27, no. 1 (Fall 2010); Volti, *Society and Technological Change*, 345–46; Bruce Bimber and David Guston, "Politics By the Same Means: Government and Science in the United States," in *Handbook of Science and Technology Studies*, chap. 24.

22. Jathan Sadowski, "The Much-Needed and Sane Congressional Office That Gingrich Killed Off and We Need Back," *Atlantic*, October 26, 2012, http://www .theatlantic.com/technology/archive/2012/10/the-much-needed-and-sane -congressional-office-that-gingrich-killed-off-and-we-need-back/264160/#.

23. Sclove, "Reinventing Technology Assessment"; Volti, *Society and Technological Change*, 345–46; Bimber and Guston, "Politics By the Same Means."

24. Hans Jonas, *The Imperative of Responsibility: In Search of an Ethics for the Technological Age* (Chicago: University of Chicago Press, 1985).

25. The precautionary principle has three interlocking components: (1) If a given technology or procedure is suspected of posing a significant risk to society, then it should not be adopted unless its projected benefits outweigh its projected drawbacks; (2) If no scientific consensus exists about the level of risk involved, then the burden of proof lies with the promoters of the technology to show that its projected benefits outweigh its drawbacks; (3) The greater the suspected level of potential harm associated with a technology, the higher the bar should be set for scientific consensus about its potential net benefits before the technology is adopted. This three-pronged articulation of the principle is my own preferred formulation—one that incorporates certain common-sense principles from risk-benefit analysis. See Bess, *The Light-Green Society*, 228–29.

26. Even though the overwhelming majority of climate scientists agree today that this is a real phenomenon, and that it is at least partly anthropogenic in nature, they still do not have absolute certainty about how rapidly it will progress and how drastic its effects will be. Our scientific models are simply not sophisticated enough yet to offer full clarity in this domain. Nevertheless, the precautionary principle does offer solid guidance: since the projected harms caused by global warming are devastating in nature, we should take decisive action today to reverse the human-induced causes of the phenomenon. We do not have full certainty, but we do have a sufficient scientific consensus to justify sharply reducing our greenhouse gas emissions. Spencer Weart, *The Discovery of Global Warming: Revised and Expanded Edition* (Cambridge, MA: Harvard University Press, 2008); Bill McKibben, *The Global Warming Reader: A*

Century of Writing About Climate Change (New York: Penguin, 2012); Michael Mann and Lee Kump, *Dire Predictions: Understanding Global Warming — The Illustrated Guide to the Findings of the IPCC* (New York: DK Publishing, 2008).

27. I offer an extended argument along these lines in Michael Bess, *Choices under Fire: Moral Dimensions of World War II* (New York: Knopf, 2006), chap. 12.

Nineteen: Enhancing Humility

1. Italo Calvino, *La giornata d'uno scrutatore* (Oscar Mondadori, 2002 [first ed. 1963]), 77. The translation of the quotation is my own.

2. The best work on this subject is Richard Rhodes, *The Making of the Atomic Bomb* (New York: Simon & Schuster, 2012). On Leo Szilard, see Michael Bess, *Realism, Utopia, and the Mushroom Cloud: Four Activist Intellectuals and their Strategies for Peace, 1945–1989* (Chicago: University of Chicago Press, 1993); and William Lanouette and Bela Silard, *Genius in the Shadows: A Biography of Leo Szilard, the Man Behind the Bomb* (Chicago: University of Chicago Press, 1994).

3. Rhodes, *The Making of the Atomic Bomb*, 305.

4. For the original letter, see http://hypertextbook.com/eworld/einstein.shtml#first.

5. Martin Sherwin, *A World Destroyed* (New York: Vintage, 1977), 27.

6. The intense debate between proponents of reform versus proponents of revolution is a recurring trope in the history of movements for social change in the modern era. I make no claim to originality in coming down firmly here on the side of the reform advocates. One of the classic instances of this debate took place within the European Left in the late nineteenth and early twentieth centuries, pitting advocates of reform such as Eduard Bernstein and Jean Jaurès against militants such as V. I. Lenin and Rosa Luxemburg. For an overview, see Leszek Kolakowski, *Main Currents of Marxism*, 3 vols., trans. P. S. Falla (New York: Norton, 2008).

7. Not surprisingly, the scholarly literature(s) on each of these three events are gargantuan, and I will not attempt here to cite them. A good place to commence with the task of comparing them: Theda Skocpol, *States and Social Revolutions: A Comparative Analysis of France, Russia and China* (New York: Cambridge University Press, 1979); Theda Skocpol, *Social Revolutions in the Modern World* (New York: Cambridge University Press, 1994); and Carles Boix and Susan Stokes, eds., *The Oxford Handbook of Comparative Politics* (New York: Oxford University Press, 2009).

8. Here again we confront massive scholarly literatures too voluminous to cite adequately in the present work. I will content myself with listing a few books in each case, referring the reader to those books' bibliographies.

9. Gail Collins, *America's Women: 400 Years of Dolls, Drudges, Helpmates, and Heroines* (New York: William Morrow, 2007); Bonnie Smith, *Changing Lives: Women in European History Since 1700* (New York: Heath, 1988); Patricia Grimshaw, Katie Holmes, and Marilyn Lake, eds., *Women's Rights and Human Rights: International Historical Perspectives* (New York: Palgrave-Macmillan, 2001).

10. E. P. Thompson, *The Making of the English Working Class* (New York: Vintage, 1966); Daniel Walkowitz and Donna Haverty-Stacke, eds., *Rethinking US Labor History: Essays on the Working-Class Experience, 1756–2009* (New York: Bloomsbury, 2010); Kolakowski, *Main Currents of Marxism*; Charles Tilly and Lesley Wood, *Social Movements 1768–2012*, 3rd ed. (Boulder, CO: Paradigm, 2012).

11. Henry Louis Gates, *Life Upon These Shores: Looking at African American History, 1513–2008* (New York: Knopf, 2011); Herb Boyd, *Autobiography of a People: Three Centuries of African American History Told by Those Who Lived It* (New York: Anchor, 2000); Jeffrey Stewart, *1001 Things Everyone Should Know About African American History* (New York: Three Rivers, 1998).

12. Five works that insightfully address this question of historical agency and effectiveness are William Sewell, *Logics of History: Social Theory and Social Transformation* (Chicago: University of Chicago Press, 2005); James Scott, *Seeing Like a State: How Certain Schemes to Improve the Human Condition Have Failed* (New Haven, CT: Yale University Press, 1998); Robert Putnam, *Making Democracy Work: Civic Traditions in Modern Italy* (Princeton, NJ: Princeton University Press, 1994); Peter Ackerman and Jack Duvall, *A Force More Powerful: A Century of Nonviolent Conflict* (New York: Palgrave-Macmillan, 2000); and Kenneth Boulding, *Stable Peace* (Austin: University of Texas Press, 1978).

13. Parker Palmer, *The Courage to Teach: Exploring the Inner Landscape of a Teacher's Life* (Hoboken, NJ: Wiley, 2007); Ken Bain, *What the Best College Teachers Do* (Cambridge, MA: Harvard University Press, 2004); The Boyer Commission on Educating Undergraduates in the Research University, *Reinventing Undergraduate Education: A Blueprint for America's Research Universities* (2006), www.umass.edu/research/system/files/boyer_fromRussell.pdf.

14. See, for example, Stephen Hall, *Wisdom: From Philosophy to Neuroscience* (New York: Vintage, 2010); Daniel Gilbert, *Stumbling on Happiness* (New York: Knopf, 2007); Jonathan Haidt, *The Happiness Hypothesis: Finding Modern Truth in Ancient Wisdom* (New York: Basic, 2006); Thomas Merton, *The Seven Storey Mountain* (New York: Harcourt, 1948, 1998); Matthieu Ricard, *Happiness* (New York: Little, Brown, 2003).

INDEX